D1037921

NATURAL SELECTIONS

ALSO BY DAVID BARASH

Sociobiology and Behavior

The Whisperings Within: Evolution and the Origins of Human Nature

Stop Nuclear War! A Handbook
(*with Judith Eve Lipton*)

Aging: an exploration

The Caveman and the Bomb: Human Nature, Evolution and Nuclear War
(*with Judith Eve Lipton*)

The Arms Race and Nuclear War

Marmots: Social Behavior and Ecology

The Great Outdoors: Star-gazing, Mountain Climbing,
Beachcombing, and other Pleasures of Nature

Introduction to Peace Studies

The L Word: An Unapologetic, Thoroughly Biased,
Long-overdue Explication and Celebration of Liberalism

Beloved Enemies: Exploring Our Need for Opponents

Gender Gap: The Biology of Male-Female Differences
(*with Judith Eve Lipton*)

Ideas of Human Nature: From the Bhagavad Gita to Sociobiology

Approaches to Peace: a reader

The Mammal in the Mirror: Understanding Our Place in the Natural World
(*with Ilona A. Barash*)

Understanding Violence

The Myth of Monogamy: Fidelity and Infidelity in Animals and People
(*with Judith Eve Lipton*)

Revolutionary Biology: The New, Gene-centered View of Life

Peace and Conflict Studies (with Charles P. Webel).

Economics as an Evolutionary Science: From Utility to Fitness
(*with Arthur and Anna Gandolfi*)

The Survival Game: How Game Theory Explains the
Biology of Cooperation and Competition

Madame Bovary's Ovaries: A Darwinian Look at Literature
(*with Nanelle R. Barash*)

NATURAL SELECTIONS

Selfish Altruists, Honest Liars, and Other Realities of Evolution

David P. Barash

BELLEVUE LITERARY PRESS
NEW YORK

I especially want to thank my friend and agent, John Michel, for his consistently sound advice and good cheer, and Erika Goldman, doyenne of Bellevue Literary Press, for her publishing vision, initiative, wisdom, and for making the making of this book a genuine pleasure. I'm also grateful to my honors seminar students at the University of Washington, who helped hone the ideas and improve the presentation, and to Malcolm Scully and others at The Chronicle of Higher Education, where much of this material originally appeared (albeit in somewhat different form). Topper and Higgins, our newly acquired three-legged orphan kittens, provided amusement and encouraging reminders of life's resourcefulness.

—DAVID P. BARASH
Redmond, WA

First published in the United States in 2008 by
Bellevue Literary Press
New York

FOR INFORMATION ADDRESS:
Bellevue Literary Press
NYU School of Medicine
550 First Avenue
OBV 640
New York, NY 10016

This book was published with the generous support of
Bellevue Literary Press's founding donor the Arnold Simon Family Trust
and the Bernard & Irene Schwartz Foundation.

Cataloging-in-Publication Data is available from the Library of Congress

Book design and type formatting by Bernard Schleifer
Manufactured in the United States of America
ISBN 978-1-934137-05-5
FIRST EDITION
1 3 5 7 9 8 6 4 2

This book is dedicated to
Eva Louise Barash,
yoga teacher extraordinaire
and much loved daughter.

Contents

1. Seductions of Centrality 9
2. Evolutionary Design Flaws, or, Why Bad Things Have Happened to Perfectly Good Creatures (Including Ourselves) 16
3. Mainstream Misconceptions 21
4. Neither Leaps Nor Bounds 32
5. Who's in Charge Here? 42
6. Material of Mind: A Surprising Homage to B. F. Skinner 49
7. Y B Conscious? 57
8. Intelligence 64
9. Let Us Reason Together 71
10. Believing Is Seeing 80
11. Evolutionary Existentialism and the Meaning of Life 86
12. The Tyranny of the Natural 98
13. Forbidden Knowledge? 104
14. Are We Selfish Altruists? Group-Oriented Individualists? (Or What?) 111
15. Dealing with Dilemmas: Personal Gain versus Public Good 118
16. The Ugly Underside of Altruism 126
17. Why Is Violence Such a "Guy Thing"? 136
18. One and a Half Cheers . . . 148
19. Honest Liars? 155
20. What Puts the Dys in Dystopia? 162
21. Evolution's Odd Couple 173

Index 185

1

Seductions of Centrality

SIGMUND FREUD WAS NOT A HUMBLE MAN. AND SO, IT WILL probably come as no surprise that when he chose to identify three great intellectual earthquakes, each of them body-blows to humanity's narcissism, his own contribution figured prominently: First, Freud listed replacement of the Ptolemaic, Earth-centered universe by its Copernican rival; second, Darwin's insights into the natural, biological origin of all living things, *Homo sapiens* included; and third, Freud's suggestion that much—indeed, most—of our mental activity goes on "underground," in the unconscious. (It is interesting to consider that even as he recounted a history of diminished human importance, Freud wasn't shy about his own!)

In any event, many of *Homo sapiens*'s most glorious scientific achievements, rather than expanding our self-image, have paradoxically diminished it. But despite this progression of self-administered "narcissistic injuries" (as Professor Freud would have it), a widespread feeling of centrality is nonetheless widespread, an insistence that the world somehow revolves around human beings, as a species and for most individuals. Many of us remain narcissists in this particular sense. Whereas infantile narcissism is plausible, predictable, and eventually outgrown, centrality remains fundamental—dare I say "central"?—to the way many adults think of themselves. But this doesn't make it true.

Almost by definition, we each experience our own private subjectivity, a personal relationship with the universe, in return for which it is widely assumed that the universe reciprocates, even though there is no evidence supporting this latter assumption . . . as well as considerable logic urging that it is untrue. Moreover, even as the illusion of centrality may be useful, if not necessary, to normal day-to-day functioning (in a sense, analogous to the denial of one's eventual death), seductive centrality is also responsible for a lot of foolishness and even mischief.

I have a friend who is paraplegic because of a rare viral infection in his spine. He was afflicted as a young adult, and although he has since managed to achieve a laudable life (loving marriage, devoted children, successful career), my friend remains obsessed with his illness, specifically *why* it happened to him. For decades, he has satisfied himself with this answer: He became ill "in order" to reconcile his parents to his then-fiancée, now wife. My friend's parents had disliked his bride-to-be, but she stood by him throughout his terrible illness and subsequent disability; her steadfastness gradually wore down their disapproval. I hasten to add that my friend is highly intelligent and well educated. But he remains convinced that the viruses lodged in his spine were somehow recruited as part of a cosmic conspiracy designed to assure his personal matrimonial bliss. Thus he has made sense of his life.

Next, consider the strange case of Tycho Brahe, which, on inspection, turns out to be not so strange after all. An influential Danish star-charter of the late 16th century, Tycho Brahe served as mentor to the great German astronomer and mathematician Johannes Kepler. In his own right, Brahe achieved remarkable accuracy in measuring the positions of planets as well as stars. But Brahe's greatest contribution (at least for my purpose) was one that he would doubtless prefer to leave forgotten, because Brahe's Blunder is one of those errors whose very wrongness can teach us quite a lot about ourselves and seduction of species-wide centrality.

Deep in his heart, Brahe rejected the newly proclaimed Copernican model of the universe, the heretical system that threatened to wrench the Earth from its privileged position at the center of all creation and relegate it to just one of many planets that circle the sun. But Brahe was also a careful scientist whose observations were undeniable, even as they made him uncomfortable: The five known planets of Brahe's day (Mercury, Venus, Mars, Jupiter, and Saturn) circled the sun. This much was settled. Copernicus, alas, was right, and nothing could be done about it. But Tycho Brahe, troubled of spirit yet inventive of mind, came up with a solution, a kind of strategic intellectual retreat and regrouping. It was ingenious, allowing him to accept what was irrefutably true, while still clinging stubbornly to what he cherished even more: what he wanted to be true. And so—like my friend, who, having no choice but to accept the fact of his illness, has also retained the illusion that it somehow arose in the service of his needs—Brahe proposed that whereas the five planets indeed circled the sun, that same sun and its planetary retinue obediently revolved around an immobile and central Earth!

My point is that Brahean solutions are not limited to astronomy or to my wheelchair-bound friend. They reveal a widespread human tendency: Whenever possible, and however illogical, we retain a sense that we are so important that the cosmos must have been structured with us in mind.

Some time ago, a brief newspaper article described a most improbable tragedy: A woman, driving on an interstate highway, had been instantly killed when a jar of grape jelly came crashing through her windshield. It seems that this jar, along with other supplies, had accidentally been left on the wing of a private airplane, which then took off and reached a substantial altitude before the jar slid off. The woman's family may well have wondered about the "meaning" of her death, just as my friend ponders the meaning of his illness, and so many people wonder about the meaning of their lives. There must be a reason, they are convinced, for their existence and for their most intense experiences. Just as Tycho Brahe struggled to avoid astronomical reality, they simply cannot accept this biological truth: They were "created" by the random union of their father's sperm and their mother's egg, tossed into this world quite by accident, just as someday they will be tossed out of it by a falling jelly jar . . . or by a delegation of rampaging viruses.

Centrality may also explain much resistance to the concept of evolution. Thus, according to Francis Bacon, "Man, if we look to final causes, may be regarded as the centre of the world . . . for the whole world works together in the service of man. . . . All things seem to be going about man's business and not their own." Such a perspective, although deluded, is comforting, and not uncommon. Thus, it may be that most of us put emphasis on the wrong word in the phrase "special creation," placing particular stress on *creation*, whereas in fact the key concept, and the one that modern fundamentalists find so attractive—verging on essential—is that it is supposed to be *special*. Think of the mythical, beloved grandmother, who lined up her grandchildren and hugged every one while whispering privately to each, "*You* are my favorite!" We long to be the favorite of god or nature, as a species no less than as individuals, and so, not surprisingly, we insist upon the notion of *special*-ness. The center of our own subjective universe, we insist on being its objective center as well.

In his celebrated and influential book *Natural Theology* (1803), William Paley wrote as follows about cosmic beneficence and species centrality:

> The hinges in the wings of an earwig, and the joints of its antennae, are as highly wrought, as if the Creator had had nothing else to finish. We see no signs of diminution of care by multiplication of objects, or of distraction of thought by variety. We have no reason to fear, therefore, our being forgotten, or overlooked, or neglected.

What my friend's delusion is to his personal tragedy and Brahe's Blunder is to the solar system, Paley's Palliative is to life on Earth: the seductive vanity of selective centrality. All speak eloquently about the human yearning for a special place in the cosmos.

A few decades earlier, Thomas Jefferson had reacted as follows to the discovery of mammoth bones: "Such is the economy of nature, that no instance can be produced of her having permitted any one race of animals to become extinct." The moral? Don't lose heart, fellow human beings! Just as there are thirty different species of lice that make their homes in the feathers of a single species of Amazonian parrot, each of them doubtless put there with *Homo sapiens* in mind, we can be confident that our existence is so important that we would never be ignored or abandoned. An accomplished amateur paleontologist, Jefferson remained convinced that there must be mammoths lumbering about somewhere in the unexplored arctic regions; similarly with the giant ground sloths whose bones had been discovered in Virginia, and which caused consternation to Jefferson's contemporaries.

At one point in Douglas Adams's hilarious *Hitchhiker's Guide to the Galaxy*, a sperm whale plaintively wonders "Why am I here. What is my purpose in life?" as it plummets toward the fictional planet Magrathea. This appealing but doomed creature had just been "called into existence" several miles above the planet's surface, when a nuclear missile, directed at our heroes' space ship, was inexplicably transformed into a sperm whale via an "Infinite Improbability Generator." Evolution, too, is an improbability generator, although its outcomes are considerably more finite. Here, then, is a potentially dispiriting message for *Homo sapiens*: Every human being—just as every hippo, halibut, or hemlock tree—is similarly called into existence by that particular improbability generator called natural selection, after which each of us has no more inherent purpose, no more reason for being, no more central significance to the cosmos, than Douglas Adams's naïve and ill-fated whale, whose blubber was soon to bespatter the Magrathean landscape.

In his famous discourse on the different kinds of causation, Aristotle distinguished, among other things, between "final" and "efficient" causes, the former being the goal or purpose of something, and the latter, the immediate mechanism responsible. Evolutionary biologist Douglas Futuyma has accordingly referred to the "sufficiency of efficient causes." In other words, since Darwin, it is no longer useful to ask "Why has a particular species been created?" It is not scientifically productive to assume that the huge panoply of millions of species—including every obscure soil microorganism and each parasite in every deep-sea fish—exists with regard to and somehow because of human beings. Similarly, it is no longer useful to suppose that we, as individuals, are the center of the universe, either. Jelly jars abound, and my friend was hit by one. Efficient causes are enough.

A case can be made that whereas my friend could be left to his misconception—which is, after all, not only harmless but genuinely consoling—*Homo sapiens* as a species needs to face the truth, especially since our puffed-up sense of ourselves appears to have figured prominently in the environmental insensitivity and abuse that has characterized so much of our collective history. In a now-classic manuscript published three decades ago in the journal *Science*, historian Lynn White identified "The Historical Roots of our Ecological Crisis" as residing in the Western religious tradition of separating humanity from the rest of the natural world, claiming Old Testament sanction for the view that we have been given dominion over all other things; that, in short, nature exists *for* us, and thus, it is our God-given right—even, our obligation—to abuse and exploit it. Human centrality, in such cases, is not only a personal, biological, and astronomical absurdity, it is downright destructive.

In this regard, we might take comfort from the several ecumenical movements that have begun to espouse "faith-based stewardship," intended to counter the troublesome Western theology of human centrality. The idea, in brief, is that human beings have a responsibility to care for God's creation. But even as I applaud this development, I cannot help registering a small shudder of distrust, because even so laudable an enterprise still revolves around the stubborn, persistent idea that We Are Special. In a sense, there isn't all that much difference between claiming that nature exists *for* us to exploit and urging that it exists *for* us to protect. Either way, *Homo sapiens* is presumed to occupy a privileged, central place in the cosmic scheme. Even theological stewardship takes it for granted that both we and the natural world were created for a purpose, part of which happens to involve taking care of nature.

The truth is more daunting. The natural world evolved as a result of mindless, purposeless material events, and human beings—not just as a species but each of us, as individuals—are equally without intrinsic meaning or purpose. "We find no vestige of a beginning," wrote pioneering geologist James Hutton, in 1788, "no prospect of an end." For some, the prospect is bracing; for others, bleak, if not terrifying. Pascal, gazing similarly into a vastness devoid of human meaning or purpose, wrote that "the silence of these infinite spaces frightens me."

Of course, maybe I am wrong, and Hutton too, and also Darwin, and Copernicus. Maybe Tycho Brahe and my paraplegic friend are correct and our planet—as well as our lives—are genuinely central to some cosmic design. Many people contend that they have a personal relationship with God; for all I know, maybe god reciprocates, tailoring his grace to every such individual, orchestrating each falling sparrow and granting to every human being precisely the degree of centrality that so many crave. Maybe we have a role to play, and maybe—as so many people in distress like to assure themselves—they will never be given more than they are capable of bearing. Maybe we aren't Magrathean whales after all, flopping meaninglessly in a foreign atmosphere, doomed to fall. (After all, in Douglas Adams's novel, there were two nuclear missiles, one transformed into a whale and the other into a pot of petunias, which made this observation, which might be seen as the author's tip-of-the-hat to Hindu reincarnation: "Oh no, not again!"). And maybe, even now, in some as yet undiscovered land, there are modern mastodons, joyously cavorting with giant sloths and their ilk, testimony to the unflagging concern of a deity or, at minimum, a natural design, that remains devoted to all creatures . . . especially, of course, ourselves.

But don't count on it.

2

Evolutionary Design, or, Why Bad Things Have Happened to Perfectly Good Creatures (Including Ourselves)

WHAT, THEN, CAN WE COUNT ON?

One possibility—ardently espoused by many—is that even if we aren't the literal (or even metaphoric) center of the universe, at least we are well designed, testimony to either God's beneficence or to evolution's remarkable powers.

In 1829, Francis Henry Egerton, the 8th Earl of Bridgewater, bequeathed 8,000 pounds sterling to the Royal Society of London to support the publication of works "On the Power, Wisdom, and Goodness of God, as Manifested in the Creation." The resulting Bridgewater Treatises, published between 1833 and 1840, were classic statements of "natural theology," seeking to demonstrate God's existence by examining the natural world's "perfection."

These days, biologists are often inclined to point, similarly, to the extraordinary complexity and near-perfection of living things, but as evidence of the power, wisdom, and goodness of natural selection, as manifested in evolution. Such gestures are understandable and perhaps even laudable, contributing as they do to a healthy gee-whiz appreciation of the Darwinian process and the organic world. But ironically, they are less useful than one might think, especially in distinguishing natural selection from its premier alternative (at least among the biologically illiterate): special creation, or, in its barely disguised incarnation, "intelligent design theory."

The problem is that those same wonders of perfection used by biologists to buttress their confidence in natural selection can also be used by believers in "intelligent design" as evidence for a divine designer. Fortunately, however, the two are in fact discriminable, some of the most powerful distinctions being provided not by the perfection of living things, but by their imperfection. Thus, it is worth emphasizing that even though natural selection regularly produces marvels of improbability (a living thing is, above all else, tremendously nonrandom and low-entropy), it is necessarily a blundering, imperfect, and tremendously unintelligent engineer, as compared to any purportedly omniscient and omnipotent creator. Ironically, it is the stupidity and inefficiency of evolution—its manifold design flaws—that argue most strongly for its material and wholly earthbound nature.

Natural selection is a mathematically precise process, whose outcome should be—and, for the most part, is—a remarkable array of "optimal" structures and systems. A naïve view therefore assumes that the biological world is essentially perfect and certainly highly predictable, like a carefully orchestrated geometric proof. Or like a billiard game, in which a skilled player can be expected to employ the correct angles, inertia, force, and momentum. And in fact, living things reveal some pretty fancy shooting. Specialists no less than biologically literate laypeople are therefore inclined to applaud, and rightly so.

And so it was that even David Hume—materialist and atheist—marveled at how the parts of living things "are adjusted to each other with an accuracy which ravishes into admiration all men who have ever contemplated them."

But admiration is not always warranted. Gilbert and Sullivan's Mikado sings about "letting the punishment fit the crime," gleefully announcing, for example, that the billiard sharp will be condemned to play "on a cloth untrue, with a twisted cue, and elliptical billiard balls." To a degree not generally appreciated, the organic world contains all sorts of imperfections, and as a result, shots often go awry . . . not because the laws of physics and geometry aren't valid, or because the player isn't skillful, but because even Minnesota Fats was subject to the sting of reality.

Make no mistake, evolution—and thus, nature—IS wonderful.

The smooth-running complexity of physiological systems, anatomical structures, ecological interactions, and behavioral adjustments are powerful testimony to the effectiveness of natural selection in generating highly nonrandom systems such as the near-incredible complexity of the human brain, the remarkable lock-and-key fit between organism and environment, the myriad interlocking details of how a cell reproduces itself, extracts energy from complex molecules, and so forth.

But imperfections intrude, and in many ways. For now, let's concentrate on just one dimension, and moreover, on just one species: *Homo sapiens*.

Among evolution's numerous constraints, one of the most vexing, and unavoidable, is history, the simple fact that living things have not been created *de novo*, but rather, have evolved from antecedents. If they were specially and intelligently designed in each case, there is no reason for the designer not to have chosen the optimum pattern in each case; insofar as they are constrained by their past, on the other hand, and the products of small incremental steps altogether lacking in foresight, living things are necessarily jerry-built and more than a little ramshackle. (It might be optimal if elephants could fly. After all, because of local overpopulation in increasingly threatened game parks, many elephants are undernourished, even starving, but for some reason they are unable to hover 30 feet above the ground and eat leaves currently beyond their reach. Walt Disney's Dumbo notwithstanding, the evolutionary past of today's pachyderms severely constrains their present and future.)

But I promised some human examples. Here goes.

Consider the skeleton. Now ask yourself, if you were designing the optimum exit for a fetus, would you engineer a route that passes through the narrow confines of the pelvic girdle? Add to this the tragic reality that childbirth is not only painful in our species, but downright dangerous and sometimes lethal, owing to occasional cephalopelvic disproportion (literally, the baby's head being too large for the mother's birth canal), breech presentation, and so forth. This design flaw is all the more dramatic since there is plenty of room for even the most stubbornly misoriented, large-brained fetus to be easily delivered, anywhere in that vast nonbony region below the ribs! And in fact, that is precisely what obstetricians do, when forced to perform a Cesarean section. (This seemingly obvious point was first made, I believe, by George C. Williams, whose book, *The Pony Fish's Glow*, is a superb source for diverse insights into evolutionary theory, design flaws included.)

It is notable that evolution has altogether neglected the simple, straightforward solution, which would have been for the vagina to open anywhere in the lower abdomen. Instead, it stubbornly and stupidly insisted on threading its way through the ridiculously narrow pelvic ring. Why? Because human beings are mammals, and therefore tetrapods by history. As such, our ancestors carried their spines parallel to the ground; it was only with our insistence on upright posture that the pelvic girdle had to be rotated, thereby making a tight birth-fit out of what for other mammals is nearly always an easy passage. An engineer who designed such a system from scratch would get a failing grade, but evolution didn't have the luxury of intelligent design. It had to make do with the materials available. (Admittedly, it can be argued that the dangers and discomforts of childbirth were preplanned after all, since Genesis gives us God's judgment upon Eve, that as punishment for her disobedience in Eden, "in pain you shall bring forth children." Might this imply that if Eve had only restrained herself, her vagina would have been where every woman's belly button currently resides?)

On to men. An especially awkward design flaw of the human body—male and female alike—results from the close anatomical association of the excretory and reproductive systems, a proximity attributable to a long-standing, primitive vertebrate connection, and one that isn't troubling only for those who are sexually fastidious. Thus, although there is no obvious downside to the deplorable fact that the male urethra does double-duty, carrying both semen and urine, most elderly men have occasion to regret that the prostate gland is closely applied to the bladder, so that enlargement of the former impinges awkwardly on the latter. In addition, as human testicles descended—both in evolution and in embryology—from their position inside the body cavity, the vas deferens, which connects testis to urethra, became looped around the ureter (which carries urine from kidneys to bladder), resulting in an altogether ridiculous arrangement that would never have occurred if evolution could have anticipated the problem and, like an even minimally competent structural engineer, designed male tubing to run in a direct line.

A final example, although plenty more are available: The primitive vertebrate system, still found among some of today's chordates, combined both feeding and respiration (just as excretion and reproduction used to overlap, and still do in many species). Water went in, food was filtered out, and passive diffusion sufficed for respiration. As body size increased, a separate respiratory system was added, but by piggybacking onto the preexisting digestive plumbing. By consequence, access to what became the lungs was achieved only by sharing a common anteroom with incoming food. As a result, people are vulnerable to choking. The Heimlich maneuver is a wonderful innovation, but it wouldn't be needed if evolution only had the foresight to design separate passages for food and air, instead of combining the two. But here, as in other respects, natural selection operated by small, mindless increments, without the slightest attention to any bigger picture or anything approaching a wise, benevolent overview.

It must be emphasized that the preceding does NOT constitute an argument against evolution; in fact, quite the opposite! Thus, if living things (including human beings) were the products of special creation rather than of natural selection, then the flawed nature of biological systems, including ourselves, would pose some awkward questions, to say the least. If God created "man" in his image, does this imply that He, too, has comparably ill-constructed knee joints, a poorly engineered lower back, dangerously narrow birth canal, and ridiculously ill-conceived urogenital plumbing? A novice engineer could have done better. The point is that these and other structural flaws aren't "antievolutionary" arguments at all, but rather cogent statements of the contingent, unplanned, entirely natural nature of natural selection. Evolution has had to make do with an array of constraints, including—but not limited to—those of past history.

We are profoundly imperfect, deep in our nature. And in these imperfections reside some of the best arguments for our equally profound natural-ness.

3

Mainstream Misconceptions

AN EVOLUTION-BASED VIEW OF LIFE YIELDS SOME WONDERFUL, counterintuitive insights, our biological imperfections prominent among them. At the same time, evolutionary thinking sometimes falls victim to this melancholy fact: The basic concept is so straightforward and seemingly simple that many people think they fully grasp it, but don't. When he first read *On the Origin of Species*, the brilliant biologist and subsequent Darwinian Thomas Huxley is reputed to have exclaimed "How stupid of me not to have thought of that!" Like most great ideas—and noted philosopher Daniel Dennett has called evolution by natural selection "the greatest single idea, ever"—there is indeed a core of simplicity about evolution. But there is also much room for misunderstanding.

Noting the role of the Royal Air Force in saving his country during the Battle of Britain, Winston Churchill observed that never have so many owed so much to so few. We owe a great deal—indeed, literally everything—to evolution, and yet, never have so many said and written so much about something they understand so poorly. Not that evolution is all that difficult to understand. Rather, so many people have such strong feelings about it, often connected to so many regrettable stock phrases that clear thought has often been obscured. This is especially unfortunate in the intellectual—or, more to the point, anti-intellectual—climate fostered by the Bush administration, which, for the initial eight years of the 21st century, persistently sought to trump science with ideology.

Notwithstanding a string of legal victories of science over so-called intelligent design during these years, the struggle is not likely to end soon. The following catalog of misconceptions followed by responses is therefore offered for those readers who may find themselves confronting voices whose amplitude and certainty exceed their wisdom.

"It's only a theory." Biologists often speak of "the theory of evolution," but not because evolution is a guess or mere speculation. My *Random House Dictionary* provides, among its definitions of "theory," the following: "a proposed explanation whose status is still conjectural, in contrast to well-established propositions that are regarded as reporting matters of actual fact." Someone might express a "theory" that Elvis Presley still lives, or that trailer parks attract tornadoes. The same person might also say "Biologists have a theory that human beings evolved," in which case, knowingly or not, a very different use of the word has been employed. Indeed, my dictionary also gives this definition of theory: "a more or less verified established explanation accounting for known facts or phenomena," with examples that include number theory, the theory of relativity, atomic theory, and so forth. In this sense, and this only, evolution is a theory. It is, in fact, as close to truth as any science is ever likely to get. (And, proudly situated in the old-fashioned, pre-postmodernist tradition, I assert that this is very, very close indeed.)

"Evolutionary logic is circular: the fittest are those that survive, and those that survive are the fittest. So it doesn't say anything." First, natural selection is not about survival, but reproduction: specifically, individuals and genes reproducing themselves. Survival is evolutionarily important because—and only because—it contributes to reproduction. Second, "fitness" does not determine natural selection; rather, natural selection is the unavoidable result of how "fit" something is, which is to say, how successful it is in promoting its genes. As such, fitness leads to the important prediction that natural selection favoring a particular type should result in a larger proportion of that type in future populations. This prediction has been repeatedly tested and confirmed.

"Natural selection is just a negative process; it cannot create anything new." Natural selection is only "negative" in that certain individuals and their genes fall by the evolutionary wayside in preference to others, which prosper. But evolution is not merely a question of delet-

ing those organisms that are less fit; because of mutation (which provides genetic novelty) and sexual reproduction (which combines DNA in unique ways), new genetic material is constantly being produced. And much depends on this regular generation of genetic diversity, on the world being, as the poet Louis MacNeice put it, "incorrigibly plural." In his poem "Snow" MacNeice went on to feel "the drunkenness of things being various," a variousness that is essential as the building blocks from which evolution constructs those things that we identify as highly adapted organisms, including ourselves.

But although the production of diversity is fundamentally random, the power of natural selection is that it is not simply at the mercy of haphazard events, merely eliminating the unfit. It creates novelty, because it adds a crucial process: a mechanism for "selective retention."

Imagine that instead of those imaginary monkeys creating all of Shakespeare, we just wanted a single phrase, "to be or not to be." It consists of 18 characters, including spaces, but not punctuation. Given the alphabet plus a possible blank space, we have a total of 27 possibilities for each slot. The chance that one of our hypothetical monkeys might randomly get the initial *t* is thus 1/27. The chance that it would simultaneously and randomly place an *o* in the second slot is 1/27 x 1/27 = 1/729. The chance of getting all 18 characters correct, by chance alone, is thus 1/27 times itself 17 times, which is inconceivably small. (The previous analysis, with some modification, was inspired by Richard Dawkins's superb book, *The Blind Watchmaker*.)

But what if, instead of tossing out every meaningless word and having to start afresh each time, those patterns that were promising—even by a little bit—were retained, and then randomly modified yet again, once more retaining (that is, selecting) those that were more "fit"? I started with 18 random strikes on a computer keyboard, and programmed the machine to make a few small changes—introduce some new letters—every "generation." These changes are equivalent to mutation and sexual recombination, providing regular sources of random variation on an existing theme. Next, I added the simple requirement of selectively retaining whatever most closely approximates "to be or not to be," after which the "organism" randomly varies again, with the outcome screened once more for resemblance—however slight—to the target. The result was that after a very small number of generations—

usually on the order of 30—the desired outcome was obtained.

In one test, for example, I started with "fuwl sazgh ekm fje." Discouraging, perhaps, but after several runs it had become the dimly recognizable "tubl hot nnoq ioby." And by run 22, it was "tu bep ok not ts e." And by 29, "to be ok not to bo," which even the most cynical monkey is likely to acknowledge as having just about arrived, and a whole lot more quickly than 1/27 to the 18th power would suggest.

Starting with gobbledygook, and using only random variation and selective retention, something new had been created, something so nonrandom, in fact, that it is perhaps the most famous phrase in the English language! One might object, of course, that there is a crucial added factor: human intelligence was injected into the process. For example, "to pee or not to pee" is also syntactically correct—and an appropriate query on certain occasions—but is unlikely to have echoed down the corridors of literature for 400 years.

Shakespeare presumably considered various possible alternatives to Hamlet's renowned dilemma, although he assuredly didn't puzzle through every option. Like a human chess master as opposed to IBM's chess-playing program, Deep Blue, a creative intellect takes numerous shortcuts. But this, too, is very much what natural selection does. Living things offer only a very limited subset of what is possible. Instead of a creative human intelligence rejecting certain verbal combinations over others, the environment faced by every living thing rejects certain genetic combinations over others. In arid conditions it rejects combinations that waste water, in cold conditions it rejects combinations that waste heat. Among predators it rejects combinations that are clumsy at stalking their prey, while among prey species it rejects those that are incautious or inept when it comes to avoiding their predators.

The fleet limbs of the antelope, the wings of birds, and the eyes of goshawks are all marvels of natural design, in no way inferior to the human design of Hamlet's melancholy question. And to understand how these were created, we need only understand how natural selection, *based on random building blocks*, can nonetheless generate highly non-random results.

"Evolution is no longer going on, especially in the case of human beings." Evolution happens any time there are changes in a popula-

tion's genetic makeup. The most powerful mechanism of evolutionary change is natural selection, which operates whenever some individuals leave more genetic representatives than others. So, the only way for evolution to cease would be if everyone reproduced equally; more precisely, if genes continued to replace themselves in exactly the same proportion as they currently exist. Just a moment's reflection should convince anyone that evolution is very alive, for human beings as for everything else, so long as "differential reproduction" is going on.

This doesn't mean, however, that the conditions of evolution are the same as they have been in the past. The "selective environment" for human beings, for example, has changed dramatically from the Pleistocene. Certain traits that almost certainly were strongly selected against, such as myopia or diabetes, are now neutral or at worst, only mildly negative. Human ingenuity has come up with eyeglasses and insulin, which only scratches the surface of how modern *Homo sapiens* has been modifying natural selection, and hence, its own evolution. Whereas it might have been selectively advantageous to be a good hunter, gatherer, or mastodon-avoider, now it is selectively advantageous to be able to reproduce despite strontium 90 in our bones, DDT in our fat, and, perhaps, to be positively attracted to ideologies that are unsympathetic to birth control. In any event, we have changed our own evolution, but not ended it.

"Biological evolution no longer matters, having been superseded by cultural evolution." Cultural evolution is real, and may in fact be the one sense in which human beings *do* experience Lamarckian evolution, via a kind of inheritance of acquired characteristics. Cultural evolution, like biological evolution, involves change, but instead of genes, it is based on cultural practice: language, technology, styles of clothing, ways of living, traditions including culinary, religious, military, intellectual, social, sexual, and just about everything else that human beings do outside the anatomy and physiology of their own bodies.

Cultural evolution, like biological evolution, requires variation, but instead of genetic mutations, its raw material comes from ideas and concepts, innovations of mind and matter that may ultimately be traceable—however indirectly—to humanity's DNA, but which are not a simple matter of biochemical alterations in genetic material.

Cultural evolution, like biological evolution, proceeds by selective retention of whatever works or is favored for any other reason (such as obvious efficiency, or the vagaries of fashion or even the dictates of the powerful). Most important, cultural evolution, because it is Lamarckian and can be "inherited" nongenetically and passed on to others within a single generation, is much faster than biological evolution. To some extent, human beings are like a train traveling on two tracks, but in our case, the wheels on one track (biological evolution) move slowly while those on the other (cultural evolution) move rapidly. No wonder our species feels pulled apart! But this disparity, rather than negating the impact of biological evolution, only italicizes its significance.

"Evolution acts for the good of the species." Not all misconceptions of evolution are errantly critical: some are simply wrong, while the "species-benefit" error—albeit wrong as well—at least has the merit of being, in some sense, favorable. Indeed, its positive cast is probably part of its appeal. How pleasant it is to think in terms of beneficence, and thus, gratifying to imagine that evolution is fundamentally looking out for each species!

The reality is that all sorts of things evolve: galaxies, stars, a person's thinking, a government's policies. But none of these qualify as evolution as biologists understand the term. People have evolved; a person does not. Evolution involves change, but not the change that takes place in growth, aging, changing one's clothes or even one's mind. To qualify as biological evolution, there must be a change in a population's gene pool over time. This is why only populations evolve, not individuals: because each of us is stuck with our private genetic endowment. And yet, individuals are crucial to evolution, both individual bodies and individual genes. In fact, one of the most important and useful realizations of recent years is that evolution operates most strongly at the lowest possible levels: notably that of individuals and genes, not groups or species.

Natural selection works by differential reproduction, with some individuals and their genes more successful than others. A species is the sum total of its individuals and their genes. It has no metaphysical existence of its own and, as far as can be determined, no one is looking out for the good of the species as a whole, although each component has been selected to look out for itself. Analogously, in a free-market society, individuals and corporations seek to maximize their profits; any larger-order benefit derived by the nation is simply the unintentional summed effects of all these private, enterprising activities going on at a lower level. In the world of living things, there is no one looking out for collective benefit, no equivalent to Medicare, the FBI, or the Department of Education. And indeed, it is because individuals and genes are selected to maximize their own fitness, with none of them looking out for the interest of the larger group, the overwhelming majority of species that have ever lived are now extinct! Furthermore, when a species is endangered and thus at risk of going under, there is no indication that its constituent individuals are especially inclined to deprive themselves for the good of the threatened whole, and every sign that living things do whatever it takes to promote their own success, not that of the species.

The good of the species is purely an artificial construct of human beings, who, identifying an emergent whole (for our own convenience), misguidedly assume that its component parts see things the same way. But in fact, when species benefit and individual benefit collide, the latter invariably wins.

"We've never found the 'missing link.'" Mark Twain once said that it was easy to stop smoking: he had done it hundreds of times! Similarly, there is no missing link: there are hundreds of them, or thousands, or millions. Consider two points, representing different species, one of which gave rise to the other. Think of them as connected by a line, representing the evolutionary continuity between them. Now, add a third point, more or less midway, and call it "the missing link." Having located this missing link, have you finished your task, and bridged the gap between the two points? Not at all. In fact, you have just produced at least two new "missing links." Fill in both of them and you are faced with four. Like the horizon, which constantly recedes if pursued, the discovery of transitional forms merely adds to the transitional forms not yet identified! Mathematicians say there is an infinite number of points between any two identified points on a line; presumably there are fewer than an infinite number of missing links, but the more we find, the more there are.

The foolishness of the concept "missing link" becomes clear when you consider that for there to be some sort of midpoint, one must specify the two ends. Granted that one is modern *Homo sapiens*. But what ancestral form, precisely, holds down the other end of the linked chain: an anthropoid ape, a monkey, a primate, a mammal, a primitive reptile, an early amphibian, a primordial vertebrate, a pre-Cambrian worm? The "missing link" between modern human beings and the Devonian fishes, for example, might be a predinosaurian reptile.

Nonetheless, some people are truly bothered by what they see as the paucity of transitional forms in the fossil record. They might just as well, however, be impressed with how many have been found. This applies to forms ancestral to *Homo sapiens* just as to other species. Probably the closest to a "missing link" in the human evolutionary lineage is the famous fossil "Lucy," a female Australopithecine ("southern ape") of the species *Australopithecus afarensis*, who stood about three feet tall and weighed around 66 pounds. Lucy is as much an intermediate between apes and people as can be imagined. But she isn't alone.

There are several other species of *Australopithecus*, some relatively slender and most if not all of them on the line that gave rise to *Homo sapiens*. Others are heavy bodied and—with the 20/20 vision of hindsight—identifiable as evolutionary dead ends whose descendants eventually went extinct. There is also a growing list of species belonging to the genus *Homo*, including *Homo habilis*, which is pretty much a link between Lucy and us, just as Lucy is a link between ancient apes and modern human beings. Other found links include *Homo erectus*, remains of which are known from Asia as well as Europe. My purpose here is not to provide a detailed list of fossil prehumans, their dates, cranial capacities, or precise relationships—a Sisyphean task at any rate because new ones are constantly turning up; rather, it is to note the existence of many links between *Homo sapiens* and earlier, ancestral animals.

While it is alive, there is no way to identify a transitional form as such. Maybe its descendants will remain largely unchanged for millions of years, so that it is not transiting to anything else, but rather, is just something that evolved early and persisted late. Or maybe its descendants will go extinct, in which case it is transitional . . . to a dead end. Or maybe its descendants will be somehow recognizable in the present day, in which case it is transitional in the usual sense of the term. In any event, rather than missing, links are actually quite abundant.

"Disputes among evolutionary biologists show that the foundations of the enterprise are shaky." Exactly the opposite is true: Creative ferment is the stuff of science. Unlike theology, with which creationists are more familiar, science is founded on ideas, discovery, testing, and refinement rather than on presumably unerring doctrines of faith. Disputes about the details of evolutionary fine tuning, far from undermining the validity of evolution, are testimony to the vitality of the whole enterprise, since any worthwhile science raises more questions than it answers. Accordingly, there is uncertainty as to whether evolution always proceeds gradually, or is punctuated by occasional bursts of change, but no question that it proceeds and that it does so by the accumulation of genetic modifications. There is also debate as to the importance of random, nonadaptive processes. (Analogously, there are disagreements among physicists about the details surrounding subatomic particles, but no dispute that such particles exist.)

"Biologists have never actually witnessed natural selection causing an evolutionary change, so the whole enterprise is therefore conjectural." Wrong again. Evolution is slow, usually taking many thousands of generations. This is not surprising, since it takes time to accumulate an observable effect when, for example, a certain gene may enjoy an advantage of only one in a thousand over its fellows. Given enough time, such a "selection differential" will make a genuine difference, but no biologist has ever lived long enough to detect significant evolutionary change in sequoia trees, for instance, or blue whales, which take many years to produce even a single generation. Hence, it is difficult to catch evolution *in flagrante delicto*. But not impossible.

Famous cases involve so-called industrial melanism among English peppered moths, the evolution of antibiotic resistance among bacteria, and observations by Peter and Rosemary Grant of adaptive changes in beak shape among Galapagos finches as a result of ongoing climate change. Admittedly, we have yet to observe one species evolving into another, but this is simply because evolutionary change is slow compared to human life spans. And besides, there is nothing magical about one species turning into another: Under the influence of artificial selection, people have caused Saint Bernards and Chihuahuas, greyhounds and bulldogs to evolve.

The key point for our purposes is that once again natural selection has been shown to give rise to evolutionary change during a very short time frame. Of course, peppered moths, finch beaks, and even antibiotic resistance may appear to be much ado about nothing, cases of evolution laboring mightily and then bringing forth mere trivialities. Such examples might seem a far cry from horses evolving from terrier-sized *Hyracotherium* to modern Budweiser behemoths, velociraptors arising from the swampy slime, or the human brain expanding from shrewlike insignificance to the crowning, cerebral glory of modern sapient humanity. All these have indeed happened, but to witness such major transitions directly it is necessary to consult the fossil record. Nonetheless, all evolutionary journeys (including the big ones, so-called "macro-evolution") begin with small steps, (so-called "micro-evolution") and are nothing but their accumulated consequences, over time.

Finally, consider this, which may well explain much opposition to evolutionary science, and which I present as distinct from the earlier misconceptions, since it is not a matter of scientific veracity but, rather, opinion: *"Evolution is a put-down, diminishing the special status of human beings."* Undergirding much opposition to evolution, I suspect, lies a deeper anxiety, that of acknowledging our kinship with "lower" life forms. "In an aversion to animals," wrote Walter Benjamin, "the predominant feeling is fear of being recognized by them through contact. The horror that stirs deep in man is an obscure awareness that in him something lives so akin to the animal that it might be recognized." That recognition does not need elaborate scientific backing; it is usually enough to look into a dog's eyes, which, interestingly, does not usually evoke horror.

In *Civilization and Its Discontents*, Sigmund Freud refers approvingly to a 19th-century German playwright, Christian Garber, who gave this advice to a would-be suicide: "We cannot fall out of this world. We are in it once and for all." This caution applies to us all: we'll eventually die, but aside from that, the world is irretrievably with us. We are stuck in the muck and glory of it all, living creatures among many, biological to the core, created by our biology no less than is a dandelion or a dolphin. We cannot fall out of it, nor is there any reason to do so. For Darwin, there was "grandeur in this view of life," in which all living things are linked both by historical continuity (that is, common ancestry) and as the products of the same fundamental process: evolution. And this is no misconception.

4

Neither Leaps Nor Bounds

It has been said that for every complex question there is an answer that is simple, satisfying, and . . . wrong. One of the simplest and most satisfying ways of understanding the world is to see it as composed of dramatic, clear-cut distinctions. Sure enough, this is usually wrong.

And so we come to one of the more interesting misconceptions of evolution, or rather, of biological processes more generally: what might be called the Fallacy of Discontinuity.

People seem to gravitate toward dichotomous propositions: up/down, in/out, black/white, good/bad, right/wrong, cowboys/Indians, the saved *versus* the damned, God *versus* the Devil, "you're either with us or you're with the terrorists," and so forth. Shades of gray, gradual transitions, imperceptible gradations, subtle shifts, delicate interpenetrations: all these, by contrast, are ungratifying, not least because they lack a kind of clarity (moral or otherwise). But this doesn't make them any less true. The natural world, in particular, is characterized by precisely these latter denizens of corporeal ambiguity, whether we like it or not. However much we yearn for clarifying, clean-cut boundaries, nature only rarely obliges, which has given rise to the old saw—variously attributed, as with the calculus, to both Newton and Leibniz—that *natura non facit saltum* ("nature does not make leaps"). Like many old saws, this one still has a few sharp teeth, some of which bite very close to home.

Take, for example, two of the hottest controversies in biomedical ethics: over abortion and stem cell research. Both disputes revolve fundamentally around a seemingly straightforward question: When does life begin? And the simple, satisfying, and wrong answer is: at conception. In turn, this issue demands attention because it relates to what is for many people another concept, seemingly straightforward but in fact nothing of the kind: that of the soul. Under this key theological construction, human beings are graced with this "something," a spark of the divine, or at least some sort of chip off the Creator's old block, eternal and sublime, distinguishing us from animals and also, presumably—unless one accepts reincarnation—something that pops into existence at some point in the ontogeny of every human being. (Either souls occupy other bodies and jump into a new one when their old abode casts off its mortal coil, or swarms of them linger about—the ethereal homeless—waiting for a suitable embodiment to present itself, or they appear *de novo* along with every new person.) But when? There is one obvious answer, satisfying to those who like to think that into each life a little leap must befall—a small step for "mankind," a big jump for the individual thereby blessed—exactly once for each of us: the instant of conception, that magical moment of "ensoulment."

By this logic, the beneficiary of such a leap is suddenly thereby rendered human, so that killing him or her is effectively murder. Hence, rigid opposition to abortion. And to stem cell research, since every one of those little cells to be experimented upon is presumed a tiny, soul-possessing human being. After all, we don't define humanity as those individuals who possess a requisite number of eyes, arms, legs, kidneys, or by the shape or function of any identifiable organ. Even anencephalics or the severely retarded are considered human because each supposedly possesses, in equal measure, a human soul.

There wriggles, however, a big fly—among many—in this theological, saltatory ointment: there is no moment of conception. In what follows, try to pick out precisely when a person becomes personified. One particular egg and one sperm, each destined to contribute one-half the genome of a future human being, is produced—along with others, doomed to be less fortunate—via complex processes of oogenesis and spermatogenesis, respectively. (Now?) The fated sperm cell must migrate through a layer of follicle cells before reaching the egg's extracellular matrix, known as the zona pellucida. The latter consists of three different glycoproteins, one of which acts as a sperm receptor, which binds to its complement on the sperm's head. (Now?) This induces a vesicle at the tip of the sperm, the acrosome, to spill its contents of hydrolytic enzymes, which enable the sperm to penetrate the zona and bump up snugly against the egg's plasma membrane. (Now?) A protein in the sperm's membrane then binds to and fuses with the egg membrane. (Now?) This in turn triggers depolarization of the latter, which prevents other sperm from entering, thereby ensuring that in the short term, only one sperm—the one in question—will do the fertilizing. (Now?) Shortly thereafter, granules in the egg's cortex release enzymes that catalyze additional, long-lasting changes in the zona, achieving a more long-lasting block to polyspermy. (Now?) Microvilli—pseudopod-like extensions of the egg's interior—proceed to transport the sperm into the egg. (Now?) Note that the two haploid nuclei—of sperm and egg—do not immediately fuse, at least not in mammals such as ourselves. Rather, the nuclear envelopes remain distinct, although they share the same spindle apparatus, through the "fertilized" egg's first mitotic division. (Now?) Only at this point, after this first division, with two daughter cells already in existence, do the parental chromosomes unite in common diploid nuclei. (Now?) But even here, the parental genes remain identifiable and distinct, as either paternally or maternally derived. (Now?) Paternal and maternal genes thus remain separate for at least 24 hours, and it takes an additional day or so before their combined influence directs cell function. There is, to repeat, no cymbal-crashing "moment" of fertilization. *Natura non facit saltum.*

This is probably just as well, because if every fertilized egg (however defined) is a person, we would be morally obliged to redirect all medical activities—and most of our other efforts—to interrupt, at all costs and by whatever means, the millions of ongoing, daily spontaneous abortions, which constitute nothing less than an immense silent holocaust. We would also, of course, have to outlaw all stem cell research, with its promise of eventually treating if not curing diabetes, spinal cord injuries, Parkinson's and Alzheimer's disease, among others.

Although the problem of ensoulment is especially dramatic, comparable difficulties arise if we substitute "mind" for "soul," since the former unquestionably derives from brain activity and the brain, too, does not arrive in a sudden flash of neuronal incandescence, to be suddenly plugged in with its complex operating system ready to start humming. A two-cell zygote has no neurons, and certainly no brain. Neither does its four-cell, eight-cell, or 128-cell descendant. Somewhere along the line, however, insensibly between egg and baby, brain cells aggregate and start whispering electrochemically to each other, whereupon a mind gradually coalesces. As to defining personhood via "viability," let's acknowledge that fetal prospects for *ex utero* survival depend on constantly changing neonatal technology, not to mention additional, gradual transitions: If a fetus is viable at seven months, what about six months and 30 days?

Similar problems bedevil the question of euthanasia, since the end of life is no more straightforward than its beginning. People can remain "alive," but in a persistent vegetative state, for literally decades. There are documented cases of others—immersed in icy water and presumed drowned—who have stopped breathing and had no discernible heartbeat for tens of minutes, and then recovered. (As a result, emergency medical responders are now trained to insist that before giving up their resuscitation efforts, a victim must be not just "dead," but "warm and dead.") Even male-female differences, which seem among the clearest examples of biologically distinct alternatives, are confounded by individuals who are transgendered, transsexual, bisexual, hermaphroditic, and so on.

The moral of all this: Natural boundaries won't ease our moral quandaries. When we most want them, they aren't there. As bioethicist Ronald Green has pointed out, we had better give up trying to *find* such boundaries, and work instead to *choose* those we can live with.

Natura non facit saltum also complicates the most persistent efforts to draw a bright line between human beings and other animals, which in turn brings in yet another key bioethical dilemma: animal rights. After all, the primary message of evolution is continuity, with species evolving from other species, nearly always via tiny, biologically imperceptible steps as individuals give rise to descendants that differ, ever so slightly, from their parents, and also from each other, in their success in reproducing. It is widely known, for instance, that *Homo sapiens* shares more than 98 percent of its genome with its closest relatives, chimpanzees (*Pan troglodytes*) and bonobos (*Pan paniscus*), which has led Jared Diamond to suggest that human beings should be designated the "third chimpanzee." With the fences down, where does one draw the line, any line? And so, Bertrand Russell found himself wondering how a resolute evolutionist could resist a proposition in favor of "votes to oysters."

Given his heartfelt endorsement of "free love," Russell would doubtless have been delighted had he lived long enough to learn that geneticists studying human and chimpanzee DNA have recently concluded that a few million years ago, pre-humans and pre-chimps produced hybrids. Of course, the very idea of ancestral human beings and chimpanzees "exchanging genes" makes many people squirm, because (let's face it) this means sexual intercourse between our ancient human and animal ancestors. It is hard enough to contemplate our parents copulating; to think of our great-great-grandparents not only descended from "monkeys," but having sex with them, too, is difficult to conceive. But conceive is what they evidently did.

There is, however, an even greater source of discomfort at work here; not simple squeamishness about sex, but a deeper repugnance that gets to the heart of why so many Americans continue to be so resistant to basic evolutionary reality. And this is why I not only welcome the news that human and chimpanzee commingled genes in the past, but I also look forward to the possibility that thanks to advances in reproductive technology, there will be hybrids, or some other mixed human-animal genetic composite, in our future.

This may seem perverse, since even the most liberal ethicists shy away from advocating the "creation" of half-person/half-animals. Why, then, am I rooting for it?

Because a powerful dose of biological reality and a further demonstration of the Fallacy of Discontinuity would be healthy indeed. And this is precisely the message that chimeras, hybrids, or mixed-species clones—"humanzees"?—would drive home.

The latest tactic of creationists in the United States has been to accept "micro-evolutionary" events, such as drug resistance in bacteria, but to draw the line at the emergence of human beings from other, "lower" life forms, cloaking their religious agenda in a miasma of pseudo-science. It is a line that exists only in the minds of those who proclaim that the human species, unlike all others, possess a spark of the divine, and that we therefore stand outside nature.

Should geneticists and developmental biologists succeed once again in joining human and nonhuman animals in a viable organism—as our ancient human and chimp forebears appear to have done long ago—it would be difficult and perhaps impossible for the special pleaders to maintain the fallacy that *Homo sapiens* is uniquely disconnected from the rest of life.

It is one thing to ignore the fact that we share roughly 98 percent of our genotype with chimpanzees; but such *ignore*-ance would require even more intellectual sleight of hand when human and nonhuman cells are literally conjoined.

Moreover, the benefits of such a physical demonstration of human-nonhuman unity would go beyond simply discomfiting the naysayers, beyond merely bolstering a "reality-based" as opposed to a bogus, "faith-based" worldview. I am thinking of the powerful payoff that would come from puncturing the most hurtful myth of all time, that of discontinuity between human beings and other life-forms. This myth is at the root of our environmental destruction, and hence, perhaps, our self-destruction.

A literal reading of Genesis suggests that human beings are not only commanded to go forth and multiply, but also to dominate and, whenever inclined, to destroy the animate world, which, lacking our unique spiritual essence, existed only for human use and abuse. Whereas "we" are special, chips off the old divine block, "they" (all other life-forms) are wholly different, made merely of matter. Hence, they don't really matter.

So let's hear it for our barrier-busting, hybridizing past as well as our future, anything that promises to drive a stake through the Fallacy of Discontinuity, and wake up *Homo sapiens* to its connection to the rest of life.

But aren't human beings unique? Of course they are. There is nothing unique, however, about being unique: Every species is! All this boundary-lessness is nonetheless horrifying to those who carry a particular brief for human uniqueness, and thus, the threat of *natura non facit saltum* has generated a continuing scramble to reconceptualize human specialness whenever other animals (and not just chimps) are discovered to possess traits previously unattributed to them: tool-use, toolmaking, complex communication, cultural traditions, laughter, the female orgasm, etc. When Jane Goodall discovered her subjects making stick tools that were then used to extract termites, her mentor, Louis Leakey, famously noted that it would now be necessary to redefine tools, or humans, or chimps!

Resistance to such redefinition seems to be driving much of the energy behind today's "intelligent design" movement, whose practitioners are in fact much less concerned with how a natural, biological process could combine different complex and seemingly independent components into one smoothly functioning process than with how they might maintain a strict, unbridgeable boundary between human beings and everything else.

Advocates of "intelligent design" by and large accept the action of evolution by natural selection when it comes to small-scale biological events, such as the evolution of drug resistance in bacteria, or successful plant and animal breeding programs. But they draw the line at any genuine kinship via transitions, maintaining that we, for example, could not possibly have evolved, by tiny, infinitesimal steps, from other primates, and thus be directly connected to the rest of life. The horror of such recognition makes anathema any direct, gradualist continuity between us and "them"; the chasm between human and animal must have always been unbridgeable, except, of course, during those days of Genesis when *Deus* deigned to *facit saltum*.

By contrast, when he developed the theory of natural selection, Darwin was crucially influenced by naturalistic gradualism, especially Charles Lyell's geological principle of uniformitarianism: that in seeking to explain the origins of natural phenomena, we must whenever possible avoid recourse to catastrophism and other imagined saltations. Instead, science should look to the evidence of processes known to have occurred or currently under way. Most of the time, these happen gradually and in very small steps.

"Natural selection," Darwin noted, "is daily and hourly scrutinizing, throughout the world, every variation, even the slightest. . . . We see nothing of these slow changes in progress, until the hand of time has marked the long lapse of ages." Evolution accordingly proceeds, he wrote, "by the preservation and accumulation of infinitesimally small inherited modifications, each profitable to the preserved being; and as modern geology has almost banished such views as the excavation of a great valley by a single diluvial wave, so will natural selection . . . banish the belief in the continued creation of new organic beings, or of any great and sudden modification in their structure." Moreover, much of Darwin's work—even when not specifically concerned with evaluating the evidence for natural selection—was concerned with gradualism for its own sake. His first major volume dealt with the (gradual) formation of reefs by the accumulated effects of tiny coral animals. He also wrote a major treatise on earthworms and how they crucially—and gradually—transform natural topography. Darwin's book on the expression of the emotions in animals and people endeavored to show an "insensible" gradation between the former and the latter.

Nonetheless, the course of gradualist thinking has not always run smoothly. There have been apostates, including even some notable biologists. William Bateson (1861-1926), for instance, was much taken with discontinuous, "meristic" variation, such as the case of longicorn Prionid beetles, which have 12-jointed antennae instead of the more usual 11 joints (he also coined the word "genetics," by the way). Hugo de Vries (1848-1935) developed a "mutation theory," based on his contention that evening primroses spontaneously gave rise to new species that "came into existence at once, fully equipped, without preparation or intermediate steps. No series of generations, no selection, no struggle for existence was needed. It was a sudden leap into another "type." Best known of this group, however, was probably Richard Goldschmidt (1878-1958), whose phrase "hopeful monster" memorably depicted the hypothesized situation of saltational, "systemic" mutants, but also quickly became a term of derision.

Goldschmidt, like Lamarck, is known (insofar as he is known at all in these ahistorical times) as someone who was wrong. Lamarck was wrong about the inheritance of acquired characteristics and Goldschmidt, about hopeful monsters. It is now widely accepted—thanks especially to the pioneering theoretical work of R. A. Fisher—that mutations in a highly adapted, closely integrated biological system are likely to *decrease* fitness in direct proportion as they are large. (A mutation, after all, is a random event: Imagine that you periodically had to rummage about in the guts of your computer, randomly disconnecting elements and adding others. What kind of intervention is more likely to cause your system to crash, one conducted with a tiny needle or a sledgehammer?) Nearly all monsters are hopeless.

Interestingly, despite the fact that both Lamarck and Goldschmidt were wrong, their ideas nonetheless appeal to the untrained: Lamarckism because of our intuitive sense of the cumulative effects of use and disuse, and Goldschmidt's legacy via popular imagery whereby the mutant survivors of a nuclear holocaust, for example, are often pictured as having two heads or a cyclopean eye, and so forth. Moreover, shortly after the rediscovery of Mendelian, particulate inheritance, natural selection actually was banished to the biological shadows, since gene-based mutations were seen as representing discontinuous leaps as opposed to the gradualism demanded by Darwinian evolution. It then fell to the great trio of R. A. Fisher, J. B. S. Haldane, and Sewall Wright to establish the "modern synthesis" by showing how the accumulation of numerous small mutational events produces gradual, continuous change of the sort demanded by natural selection and widely observed in reality.

Recently, the most prominent opponent of *natura non facit saltum* was doubtless Stephen Jay Gould, who, along with Niles Eldridge, propounded the notions of "punctuated equilibrium" and "species selection." Evolutionary change often appears abrupt and discontinuous as viewed in the fossil record, but as Darwin emphasized, that record is itself incomplete and discontinuous in the extreme. But this doesn't mean that life is.

At least some extinctions have indeed been extraordinarily quick and devastatingly complete, such as the great late-Cretaceous dinosaur die-off, an apparent result of asteroid impacts. But even here, the rise of mammals (and of birds and bony fish) was gradual, not abrupt, taking many thousands of generations. Nor do evolutionary changes occur at a constant rate. Sometimes they are comparatively rapid, sometimes slow, but always they involve the differential reproduction of certain minor variants over others. Moreover, whereas natural events have occasionally been effective in ending—almost instantaneously—the reign of certain organic beings, it has not acted similarly when it comes to replacing them. Even the so-called Cambrian explosion took many millions of years. As would-be nation-builders have been learning in Afghanistan and Iraq, it is easier to destroy than to build. Extinction can sometimes be (at least somewhat) saltatory; "creation" can't.

The conclusion is inescapable: Simply by being alive, we all occupy a slippery slope. One thing grades into another, such that even living/nonliving isn't a yes/no affair. We know that we are alive. But are viruses? They can be crystallized, in which state they are metabolically "dead," seemingly forever . . . until they invade a living cell, whereupon they take over its internal machinery. This is one reason why viruses are so difficult to kill; one can imagine them thumbing their noses at us, taunting "You can't kill us, we're already dead!" And what about prions, which cause, for instance, mad cow disease? They are even "deader" than viruses, being merely proteins folded in a peculiar and potentially lethal manner.

Once, it was thought, that life embodied a "vital principle," which fundamentally distinguished it from non-life. Hence the excitement (among some) and consternation (for others) occasioned by the synthesis of urea, the first organic compound produced in a laboratory. And of course, similar interest has attended efforts at pinning down exactly how, about four billion years ago, certain organic molecules became self-replicating, thereby crossing the fuzzy line that distinguished the "living" from "non-living" inhabitants of the Earth's long-simmering organic soup.

Whether at the beginning of the evolution of life itself, or the beginning—and end—of each individual life, *natura non facit saltum*. We can look to nature for clear boundaries, major discontinuous jumps, and easy answers. But since we are unlikely to find them, it is incumbent upon us to make choices. Nature is far more likely to roll than to bounce, which leaves the ball in our court.

Let's leave the last words to a modern icon of organic, oceanic wisdom, SpongeBob SquarePants, a cartoon character recognizable by many children and more than a few adults. Mr. SquarePants, a cheerful, talkative—although admittedly, somewhat cartoonish—fellow of the phylum Porifera, is described in his theme song as being "absorbent," "yellow," and "porous." I don't know about the yellow, but absorbent and porous are we, too.

5

Who's in Charge Here?

If, as suggested at the end of the last chapter, we are absorbent and porous, continuous with the rest of life rather than rigidly separated from it, the following question arises: Who, or what, is that "we," or "you," or "me"? Another way of saying this: Who's calling the shots? When big corporations misbehave, people ask whether the CEO has been in the driver's seat all along, or was he merely a clueless figurehead, manipulated by a nefarious CFO?

It's a different story, however, when we consider ourselves. Everyone knows the Chief Executive Officer inside his or her head, where—no politics or legalism involved—the buck stops. John Donne be damned: Each person is an island, entire of him- or herself, everyone his own man or woman, separate, distinct, independent, and in charge. An army of one. This at least is what cultural tradition and subjective experience tell us.

But wait a moment. Think of the morgue scene in the movie *Men in Black*, when what appears to be a human corpse is dissected and revealed to be a highly realistic robot, its skull inhabited by a little green man from outer space. Or for a less fanciful example, turn to the disconcerting fact that there are many more parasitic creatures than free-living counterparts; after all, pretty much every multicellular animal is home to numerous fellow travelers, and—this is the point—each of these creatures has its own agenda. Considering just one group of worms, invertebrate biologist Ralph Buchsbaum suggested that "if all the matter in the universe except the nematodes were swept away, our world would still be dimly recognizable. . . . Trees would still stand in ghostly rows representing our streets and highways. The location of the various plants and animals would still be decipherable, and, had we sufficient knowledge, in many cases even their species could be determined by an examination of their erstwhile nematode parasites."

What difference does this make? For many of us supposedly "free-living" creatures, quite a lot. Providing room and board to other life-forms doesn't only compromise one's nutritional status (not to mention peace of mind), it often reduces freedom of action, too. The technical phrase is "host manipulation." For example, the tapeworm *Echinococcus multilocularus* causes its mouse "host" to become obese and sluggish, making it easy pickings for predators, notably foxes which —not coincidentally—constitute the next phase in the tapeworm's life cycle. Those the gods intend to bring low, according to the Greeks, they first make proud; those tapeworms intending to migrate from mouse to fox do so by first making "their" mouse fat, sluggish, and thus, fox-food.

Sometimes the process is more bizarre. For example, the life cycle of a trematode worm known as *Dicrocoelium dentriticum* involves doing time inside an ant, followed by a sheep. Getting from its insect host to its mammalian one is a bit of a stretch, but the resourceful worm has found a way: Ensconced within an ant, some of the worms migrate to its formicine brain, whereupon they manage to rewire their host's neurons and hijack its behavior. The manipulated ant, acting with zombielike fidelity to *Dicrocoelium*'s demands, climbs to the top of a blade of grass and clamps down with its jaws, whereupon it waits patiently and conspicuously until it is consumed by (you guessed it), a grazing sheep. Thus transported to its desired happy breeding ground deep inside sheep bowels, the worm turns, or rather, releases its eggs, which depart with a healthy helping of sheep poop, only to be consumed once more, by ants. It's a distressingly frequent story . . . distressing, at least, to those committed to "autonomy."

A final example, as unappetizing as it is important: Plague, the notorious Black Death, is caused by bacteria carried by fleas, which, in turn, live mostly on rats. Rat fleas sup cheerfully on rat blood, but will happily nibble people, too; and when they are infected with the plague bacillus, they spread the illness from rat to human. The important point for our purposes is that once they are infected with plague, disease-ridden fleas are especially enthusiastic diners, because the plague bacillus multiplies within flea stomachs, diabolically rendering the tiny insects incapable of satisfying their growing hunger. Not only are these fleas especially voracious in their frustration, but because bacteria are cramming its own belly, an infected flea vomits blood back into the wound it has just made, introducing plague bacilli into yet another victim. A desperately hungry, frustrated, plague-promoting flea, if asked, might well claim that "the devil made me do it," but in fact, it is the handiwork of *Pasteurellis pestis*. ("So, naturalists observe, a flea has smaller fleas that on him prey," wrote Jonathan Swift. "And these have smaller still to bite 'em. And so proceed ad infinitum.")

Not that a plague bacterium—any more than a mouse-dwelling tapeworm or ant-hijacking "brain-worm"—knows what it is doing when it reorders the inclinations of its host. Rather, a long evolutionary history has arranged things so that the manipulators have inherited the earth.

Not quite so simple, however. The ways of natural selection are devious and deep, embracing not only would-be manipulators but also their intended victims. Hosts needn't meekly follow just because others seek to lead. Sometimes it is unclear whether seemingly free spirits act at the behest of others, or themselves. Take coughing, or sneezing, or even—since we have already broached some indelicate matters—diarrhea.

When people get sick, they often cough and sneeze. Indeed, aside from feeling crummy or possibly running a fever, coughing and sneezing are important ways we identify being ill in the first place. It may be beneficial for an infected person to cough up and sneeze out some of the tiny organismic invaders, although to be sure, it isn't so beneficial for others nearby. This, in turn, leads to an interesting possibility: What if coughing and sneezing aren't merely symptoms of disease, but also—even primarily—a manipulation of us, the "host," by, say, influenza virus? Shades of fattened mice and grass blade–besotted ants. As to diarrhea, consider that it is a major (and potentially deadly) consequence of cholera. To be sure, as with a flu victim's sneezing and coughing, perhaps it benefits a cholera sufferer to expel the cholera-causing critter, *Vibrio cholerae*. But—and here is the key point—

it also benefits *Vibrio cholerae*. Just as Lenin urged us to ask "who, whom?" with regard to social interactions—who benefits at the expense of whom?—an evolutionary perspective urges upon us the wisdom of asking a similar question. Who benefits when a cholera victim "shits his guts out" and dies? Answer: the cholera bacillus.

This most dramatic symptom of cholera is caused by a toxin produced by the bacillus, making the host's intestines permeable to water, which gushes into the gut in vast quantities. This produces a colonic flood that washes out much of the native bacterial intestinal flora, leaving a comparatively competitor-free environment in which *Vibrio cholerae* can flourish. A big part of that "flourishing," moreover, occurs via flushing, with more than 100 billion *V. cholerae* per liter of effluent sluicing out of the victim's body, whereupon, if conditions are less than hygienic, they can infect new victims. Diarrhea, then, isn't just a symptom of cholera, it is a successful manipulation of *Homo sapiens*, by the bacteria and for the bacteria.

Intriguing as these tales of pathology may be, it is too easy to shrug them off when it comes to the daily, undiseased lives most of us experience. After all, aside from sneezing, coughing, or pooping, our actions are, we like to insist, ours and ours alone, if only because we are acting on our own volition and not for the benefit of some parasitic or pathogenic occupying army. So when we fall in love, we do so for ourselves, not at the behest of a romance-addled tapeworm. When we help a friend, we aren't being manipulated by an altruistic bacterium. If we eat when hungry, sleep when tired, scratch an itch, or write a poem, we aren't knuckling under to the needs of our nematodes. But in fact, it isn't that simple.

Think about having a child. I don't mean think about how it feels, what it costs, or the social, family, personal, or cultural factors involved. Rather, think about it as Lenin suggested (who, whom?), or as a modern-day Darwinian might: Who—or rather, what—benefits from reproduction? It's the genes, stupid. They're the beneficiaries of baby-making, the reason for reproducing. As modern evolutionary biologists increasingly recognize, bodies—more to the point, babies—are our genes' way of projecting themselves into the future. Bodies are temporary, ephemeral, short-lived survival vehicles for those genes, which are the only entities that persist over evolutionary time.

No matter how much money, time, or effort is lavished on them, regardless of how much they are exercised, pampered, or monitored for bad cholesterol, bodies don't have much of a future. In the scheme of things, they are as ephemeral as a spring day, a flower's petal, a gust of wind. What does persist is not bodies, but *genes*. Bodies go the way of all flesh: ashes to ashes and dust to dust, molecule to molecule, and atom to atom. Bodies are temporary, cobbled together from recycled parts scavenged from the cosmic junk-heap. *Genes*, on the other hand, are potentially immortal.

In his poem "Heredity," Thomas Hardy had a premonition of modern evolutionary biology and the endurance of genes:

> I am the family face
> Flesh perishes, I live on
> Projecting trait and trace
> Through time to times anon,
> And leaping from place to place
> Over oblivion.
>
> The years-heired feature that can
> In curve and voice and eye
> Despise the human span
> Of durance—that is I;
> The eternal thing in man,
> That heeds no call to die.

More troublesome, for people worried about the question—"Who's in charge here?"—that opened this meditation: Who's calling and who's heeding? The biologically informed answer is not all that different from those alarming rat/tapeworm, ant/trematode, flea/bacteria relationships, only this time it's genes/body. Unlike the cases of parasites or pathogens, when it comes to genes manipulating "their" bodies, the situation seems less dire to contemplate, if only because it is less a matter of demonic possession than of *our* genes, *our*selves. The problem, however, is that those presumably personal genes aren't any more hesitant about manipulating our selves than is a brain-worm hijacking an ant.

Take a seemingly more benign behavior, indeed, one that is highly esteemed: altruism. This is a favorite of evolutionary biologists, because superficially, every altruistic act is a paradox. Natural selection should squash any genetically mediated tendency to confer benefits on someone else while disadvantaging the altruist. Such genes should disappear from the gene pool, to be replaced by their more selfish counterparts. To a large extent, however, the paradox of altruism has been resolved by the recognition that "selfish genes" can promote themselves (rather, identical copies of themselves) by conferring benefits on genetic relatives, who are likely to carry copies of the genes in question. By this process, known as "kin selection," behavior that appears altruistic at the level of bodies is revealed to be selfish at the level of genes. Nepotism is natural. (So, by the way, is gangrene; natural and "good" aren't necessarily the same.) When someone favors a genetic relative, who, then, is doing the favoring: the nepotist or the nepotist's genes?

Just as sneezing may well be a successful manipulation of "us" (*Homo sapiens*) by "them" (viruses), what about altruism as another successful manipulation of "us," this time by our own "altruism genes"? Admirable as altruism may be, it is therefore, in a sense, yet another form of manipulation, with the manipulated victim (the altruist) acting at the behest of some of his or her own genes. After all, just as the brain-worm gains by orchestrating the actions of an ant, altruism genes stand to gain when we are nice to cousin Sarah, never mind that such niceness is costly for the helper.

All this may seem a bit naïve, since biologists know that genes don't order their bodies around. No characteristic of any living thing emerges full grown from the coils of DNA, like Athena leaping out of the forehead of Zeus. Every trait—including behavior—results from a complex interaction of genetic potential and experience, learning as well as instinct, nurture inextricably combined with nature. Life is a matter of genetic influence, not determinism.

But does this really resolve the problem? Let's say that a brain-worm–bearing ant still possesses some free will. And that a trematode-carrying mouse has even a bit more. So what if their behavior were influenced, but not determined? Wouldn't even "influence" be enough to cast doubt on their agency, their independence of action?

And (here comes the Big Question), why should human agency or free will be any less suspect? Even if we are manipulated just an eents-y weentsy bit by our genes, isn't that enough to raise once again that disconcerting question: Who's in charge here?

Maybe it doesn't matter. Or, put differently, maybe there is no one in charge. That is, no one distinguishable from everyone and everything else. If so, then this is because the "environment" is no more outside us than inside, part tapeworm, part bacterium, part genes, and no independent, self-serving, order-issuing homunculus. Purveyors of Buddhist wisdom note that our skin doesn't separate our organismal selves from the environment.It joins us to it, just as purveyors of biological wisdom know that we are manipulated by the rest of life no less than we are manipulators of the rest of life. And so, once again: Who's in charge here? Well, that depends on what the meaning of "who" is. Who's left after the parasites and pathogens are removed? And after "you" are separated from your genes?

6

Material of Mind: A Surprising Homage to B. F. Skinner

HERE ARE THE OPENING LINES OF FRANCIS CRICK'S IMPORTANT book, *The Astonishing Hypothesis*: "You, your joys and your sorrows, your memories and your ambitions, your sense of personal identity and free will, are in fact no more than the behavior of a vast assembly of nerve cells and their associated molecules. As Lewis Carroll's Alice might have phrased it, 'You're nothing but a pack of neurons.'"

In his later years, Crick, one of the towering biologists of the 20th century, turned his attention from unraveling the structure of genes (he was arguably the more creative, and certainly, the more pleasant and entertaining part of the famous Watson and Crick Nobel Prize–winning team), to looking at consciousness—not as a metaphysical phenomenon but a biological one. There is a stunningly stubborn and, to my thinking, altogether admirable materialism at the heart of evolutionary biology, even as it questions the very existence of a "self" independent of other organisms, not to mention our own genes. And so, it is consistent that a thoroughly biological cast of mind leads to a thoroughly biological conception of mind itself.

Which leads me, paradoxically as it might seem, to B. F. Skinner.

I am not now, nor have I ever been, a devotee of behaviorism, radical or otherwise.

Moreover, when I teach or write about animal behavior, I often counterpoise B. F. Skinner's work in particular as the intellectual antipode of my own perspective, which emphasizes the importance of built-in, prewired, evolutionarily generated mechanisms. For Skinner and his disciples, by contrast, living things (including human beings) are *tabula rasa*, blank slates upon which the contingencies of reinforcement write as they will, thereby constituting the crucial—indeed, the only—determinant of behavior: the experience of each individual. By contrast, I think it much more likely that living things (including, once again, human beings) are palimpsests, tablets that are far from blank, because natural selection has written upon them, then crossed out and rewritten, doing this again and again, innovating, erasing, revising, and correcting, passing down our "nature" as a heavily edited evolutionary bequeathal, a much overwritten tablet of DNA.

All this could hardly be more different from behaviorism, whose more hard-nosed version denies the very existence of "human nature." Consider, for instance, this famed pronouncement by John Watson, Skinner's conceptual mentor—"Give me a dozen healthy infants, well-formed, and my own specified world to bring them up in and I'll guarantee to take any one at random and train him to become any type of specialist I might select—doctor, lawyer, artist, merchant-chief and yes, even beggar-man and thief, regardless of his talents, penchants, tendencies, abilities, vocations, and race of his ancestors."

To be fair, Skinner occasionally wrote approvingly about the existence of species differences, even acknowledging the role of evolution by natural selection in generating these differences. Nonetheless, his seminal book, *The Behavior of Organisms,* is in fact about the bar-pressing behavior of white rats. (An insightful cartoon shows two rats in a Skinner box, conversing while a white lab-coated psychologist looms over them. One rat to the other: "Boy, do I have this guy trained. Every time I press the bar, he gives me food!")

Into my own comfortable conceptual dichotomy—"behaviorism bad; evolutionism good," a formulation worthy of Orwell's *Animal Farm*—there came an apple of discord, or rather, of concord between me and my previously satisfying dismissal of radical behaviorism, when I happened to reread Skinner's *Beyond Free Will and Dignity*, published more than three decades ago.

Please don't misunderstand: I haven't become a convert to behaviorism. But I have emerged with a deeper respect for B. F. Skinner and his work, and a recognition that in his legacy, not just evolutionary biologists but all scientists have a potent intellectual ally.

"Sociobiology and Skinner"? Oxymoronic indeed, although "science and Skinner" is a bit more coherent, since the driving force of Skinner's work was a passion to make the study of behavior "scientific" at last. My point is that even if we choose to discount Skinner's claim that reinforcement is *the* key to behavior—and discount it I do—there is deep wisdom in his pioneering insistence upon science as the fundamental paradigm for explaining human actions, even if the specific approach he pioneered has been found wanting. Moreover, in a time of rising religious fundamentalism, abetted by powerful political allies, as well as a revival of pseudo-sciences of all sorts—not to mention the postmodernist denial of the legitimacy of science itself—we need all the clear-eyed thinking we can get, especially when applied to our own species.

For me, in short, the issue is not whether Skinner was correct in the particular paradigm he espoused, but rather, his prescience in pushing students of behavior to embrace the broader paradigm of science, with its emphasis on objective, mechanistic explanations.

The problem is not simply one of seeing ourselves as others see us, but as we really are. Thus, for a long time the best view in the city of Warsaw has been from the top of the Palace of Culture. Why? Because this is practically the only place in that otherwise appealing city from which it is impossible to *see* the Palace of Culture (a thoroughly regrettable example of Stalinist architecture at its worst). By the same token, we all see the world from the Palace of our own perceptions, having only this very limited viewpoint from which to see ourselves.

It was Skinner who identified, more clearly than anyone before (or after), the key stumbling block for those of us trying to see ourselves accurately; namely, a reluctance to countenance that human actions are *caused*, because the more causation, the less credit. "We recognize a person's dignity or worth," writes Skinner, "when we give him credit for what he has done. The amount we give is inversely proportional to the conspicuousness of the causes of his behavior. If we do not know why a person acts as he does, we attribute his behavior to him. We try to gain additional credit for ourselves by concealing the reasons why we behave in given ways or by claiming to have acted for less powerful reasons." Ironically, there is something flattering and legitimizing in actions or thoughts that spring unbidden from our "self"—whatever that may be—and which aren't otherwise explicable. By the same token, the more our actions are caused, the less are we credited for them.

Skinner, again:

> Any evidence that a person's behavior may be attributed to external circumstances seems to threaten his dignity or worth. We are not inclined to give a person credit for achievements which are in fact due to forces over which he has no control. We tolerate a certain amount of such evidence, as we accept without alarm some evidence that a man is not free. No one is greatly disturbed when important details of works of art and literature, political careers, and scientific discoveries are attributed to "influences" in the lives of artists, writers, statesmen, and scientists respectively. But as an analysis of behavior adds further evidence, the achievements for which a person himself is to be given credit seem to approach zero, and both the evidence and the science which produces it are then challenged.

And not only achievements: the quotidian events of normal living also qualify.

Most of my students are alternately amused and troubled, for example, when I speculate that "love" is, on one level, an evolutionary mechanism that ensures an inclination to invest in individuals suitable to help maximize one's fitness, and on another, a consequence of appropriate amounts of oxytocin (in women) or vasopressin (in men), released in conjunction with sexual satisfaction. "That's just not acceptable," one young lady moaned, "I want my boyfriend to love me on his own, and not because of his genes or chemicals, but because of *him* and *me*!"

It is one thing, however, to insist on being loved for one's self, and not, for example, because of a hefty trust account; quite another to demand that love emerge spontaneously, somehow bubbling up and taking form without any cause whatsoever!

Skinner points out, further, that a scientific conception of behavior "does not dehumanize man, it dehomunculizes him," abolishing the unsupportable conceit that people are responsible for their actions. Why unsupportable? After all, each of us knows, subjectively, that we are free to act as we choose, and yet, as David Hume pointed out more than two centuries ago, this "knowledge" must be false: either our behavior is a consequence of prior events (modern readers can substitute "contingencies of reinforcement," "genetic predispositions toward fitness maximization," "electrochemical events taking place across neuronal membranes," and so forth), in which case we are not responsible for such actions, or it is truly spontaneous and thus random, in which case we are, if anything, even less responsible.

Thus are we transported to the ancient and seemingly insoluble conundrum of free will, which most of us "solve" by adopting two altogether inconsistent viewpoints. On the one hand, anyone espousing science—or even something as basic as cause and effect—cannot help acknowledging that free will must be an illusion insofar as *everything* is caused. But on the other, nearly all of us act in our daily lives as though we possess free will in abundance, and, moreover, that others do, too. Do we contradict ourselves? Very well, a modern-day Whitman might conclude, we contradict ourselves. We are large; we contain multitudes.

Skinner points with bemusement to essayist Joseph Wood Krutch's lament that humanity's self-conception has greatly deteriorated, from Hamlet's "How like a god!" to Pavlov's "How like a dog." One needn't be a behaviorist—or a Pavlovian—to conclude, however, that this transition constitutes progress. Whether god or homunculus or Wizard of Oz hiding behind the curtains with his hand on the levers of power, there is little to be gained from such metaphysical explanatory fictions; by contrast, although *Homo sapiens* are more complex than dogs, their dog-nature—unlike their purportedly divine essence—is at least amenable to scientific analysis and rational understanding. In short, from god to dog is a step up.

But isn't it demeaning? And even dangerous? After all, in 1999, on the floor of the U. S. Congress, majority leader Tom DeLay (R-Texas—subsequently driven out of politics by his corruption) blamed the shootings at Columbine High School on the teaching of evolution:

"Our school systems teach the children that they are nothing but glorified apes who are evolutionized [sic] out of some primordial soup."

On the other hand, Darwin, in the final paragraph of *Origin of Species*, suggested instead that "there is grandeur in this view of life," one that recognizes the connectedness of our species to the rest of evolution. Such connectedness requires not only historical continuity but also continuity of mechanism, at the level of organs, cells, organelles, molecules, and so forth. As to dangerous, here is Skinner, once again: "The problem is to free men, not from control, but from certain kinds of control. . . ." That is, we may choose ignorance over self-knowledge, but this will not in itself make us into autonomous creatures. We are influenced, and to some degree even controlled, by what surrounds us (as well as what emanates from our DNA). It is no coincidence that John Watson became a major figure in the nascent advertising industry.

What about diminution of our free will? Skinner takes that one on, too, when he notes that "no theory changes what it is a theory about." If we had free will before Skinner, or Darwin, or recent pioneers in neurobiology, then nothing in their work can take it away. And likewise, insofar as human behavior is already "controlled," then science will not free us. Well, actually, that's not quite true: The more we understand about the nature of whatever control already exists (at the level of "reinforcers," neurobiology, genetic predispositions, etc.), the freer we are to design the kinds of control we would like. Skinner is quite clear that the goal is not to free human behavior from control—because in his opinion, that can never be—but to introduce some choice as to the *kinds* of control. And this, paradoxically, promises to put "us," whoever that is, back into the driver's seat, or at least nearby.

Throughout his work, Skinner studiously avoided any intimation as to consciousness, subjectivity, or their underlying neural mechanisms, not because he denied their existence, but because he maintained that they could not be scientifically investigated. Not everyone has agreed. Thus, more than two thousand years ago, Lucretius argued that "the nature of the mind and soul is bodily," pointing out that the mind is clearly material since it is predictably influenced by blows from weapons or the ravishes of disease, stimulated—or depressed—by food and drink, and abolished with death. And in *Man a Machine*, physician Julien Offray de la Mettrie spoke for the 18th-century Enlightenment when he noted that "we can attribute the admirable property of thinking to matter," and that "to be a machine, to feel, think, know good from evil like blue from yellow, in a word, to be born with intelligence and a sure instinct for morality, and yet to be only an animal ('How like a dog' redux) are things no more contradictory than to be an ape or a parrot and know how to find sexual pleasure."

"Thought," wrote de la Mettrie, "Is so far from being incompatible with organized matter that it seems to me to be just another of its properties, such as electricity, the motive faculty, impenetrability, extension, etc."

To repeat, Skinner didn't concern himself with the neurobiological mechanisms of thought or consciousness, not because he denied the connection, but because—like Freud—he felt that its elucidation was far in the future. These days, given extraordinary advances in neurobiology, not to mention plain old-fashioned common sense, such a connection is beyond dispute. Scientists as well as any nonscientists with an empirical turn of mind realize that there is nothing whatsoever astonishing about Francis Crick's "hypothesis," which is overwhelmingly recognized as fact. Indeed, it would be astonishing if *not* true, if mental processes did not derive—wholly and completely—from neurons. And this, in turn, only italicizes the wisdom of Skinner's insistence that we would do well to stop deluding ourselves, and start accepting that behavior—like all other natural processes—is caused. People may aspire to dignity, or inherently possess it, or struggle to achieve it, but such dignity is not impeded by the fact that they are embodied, evolved creatures functioning in a physical world. This is a major part of Skinner's legacy.

And yet, in a different sense, Crick's book—like Skinner's—was well titled. Although it is a mundane fact, generally taken for granted among all scientists, that Descartes was wrong and there is no dualism separating mind from body, the reality of embodiment (and thus, the dependence of mind on body) is, in its own way, astonishing.

In a science fiction story by Terry Bisson, we listen in on a conversation between the robotic commander of an interplanetary expedition and his equally electronic leader, reporting with astonishment that the human inhabitants of Earth are—gasp!—"made out of meat." "Meat?" "There's no doubt about it. . . ." "That's impossible. . . . How can meat make a machine? You're asking me to believe in sentient meat." "I'm not asking you. I'm telling you. These creatures are the only sentient race in the sector and they're made out of meat." . . . "Do you have any idea of the life span of meat?" "Spare me." "Okay, maybe they're only part meat. ..." "Nope, we thought of that, since they do have meat heads . . . But . . . they're meat all the way through." "No brain?" "Oh, there is a brain all right. It's just that the brain is made out of meat!" "So . . . what does the thinking?" "You're not understanding, are you? The brain does the thinking. The meat." "Thinking meat! You're asking me to believe in thinking meat?" "Yes, thinking meat! Conscious meat! Dreaming meat! The meat is the whole deal! Are you getting the picture?"

Thanks in large part to Skinner, we are.

7

Y B Conscious?

CONSCIOUSNESS HAS LONG BEEN THE THIRD RAIL OF BIOLOGY: touch it and . . . maybe you don't die, but you are unlikely to get tenure. It helps, of course, if you are a Nobel laureate, such as Francis Crick or Gerald Edelman, but until recently it appeared that even their attempts to pin down the electrical-chemical-anatomical (or whatever) substrate of consciousness would go the way of Einstein's doomed search for a unified theory of everything. This may yet be the case, but the situation has nonetheless changed dramatically of late, such that inquiry into the neurobiology of consciousness has become one of the hottest, best-funded, and most media-attracting of research enterprises, along with genomics, stem cells, and a few other newly favored subdisciplines.

For literally centuries, it was perfectly acceptable for philosophers to ponder consciousness, because after all, no one really expected them to come up with anything real. (Descartes' renowned *cogito*, for example, was modified thusly by Ambrose Bierce: *cogito cogito ergo cogito sum*—"I think I think, therefore I think I am," to which Bierce added that this was about as close to truth as philosophy is likely to get!) But now we have microelectrodes recording from individual neurons, computer modeling of neural nets, functional MRIs, and an array of even newer 21st-century techniques, all hot on the trail of how consciousness emerges from "mere" matter. Cartesian dualism is on the run, as well it should be.

Admittedly, there are some exceptions, proving that imbecility runs deep, especially in the curious world of the consciousness-credulous. Take the remarkable popularity of the charlatan-cinema

"What the Bleep Do We Know!?" with its faux-scientific assertion that consciousness is an active force by which we can impact the world, not to mention showcasing Masaru Emoto's ludicrous—and persistently unreplicated—claim that water forms different kinds of crystals as a result of being exposed to "fields of consciousness" embodied in written messages such as "You fool" (no crystals or ugly ones) as compared to "I love you" (beautiful, heartwarming symmetrical delights). With such friends, the serious study of consciousness hardly needs enemies.

My intent, however, is neither to bury nor to praise neurobiology, but to point instead to another side of bona fide "consciousness research" that has received all too little attention. I refer to the question of why consciousness exists at all. Evolutionary biologists find it useful to distinguish two basic kinds of questions: proximate and ultimate, which essentially equates to "how" versus "why." Thus, for example, when inquiring into the causes of bird migration, we might examine the possible role of hormones, or of changes in day length, food availability, and so forth. Or look into the particular brain regions that are involved, the potential impact of social learning versus instinct, etc. All are valid approaches, but they share a common limitation: each is concerned only with the proximate mechanisms that initiate migration, matters of "how."

By contrast, inquiry into ultimate or evolutionary causation concerns itself with "why." Why migrate at all, instead of staying home? Given the costs of migration, what are the benefits that have presumably favored the evolution of this phenomenon—regardless of its attendant proximate mechanisms—in the first place? Biological research is at its best when grappling with both proximate *and* ultimate causation.

So, let's grant a "how" to consciousness: some how or other, energy and matter come together and produce it, via electrochemical events across neuronal membranes. What about the "why"? After all, it is quite possible to imagine a world inhabited by highly competent (even highly intelligent) zombies, who go about their days responding appropriately to stimuli—basking, perhaps, in the warm sun, obtaining suitable nutrients at opportune times, even repairing themselves—but lacking consciousness. Computers are highly intelligent; they can play winning chess and perform very difficult calculations, but they don't show any signs of possessing an independent and potentially even rebellious self-awareness, like HAL, in *2001: A Space Odyssey*.

Consciousness may well be a *sine qua non*, necessary but not sufficient for human-ness, all of which leads inevitably to the question: Why has consciousness evolved?

First, a brief attempt to define it—or rather, a gesture in that direction, and a modification of Potter Stewart's oft-repeated observation concerning pornography: we may not be able to define consciousness, but we know it when we experience it. I propose that consciousness can usefully be identified as a particular example of awareness (whatever that is!), characterized by a curious recursiveness in which individuals are not only aware, but aware that they are aware. By this conception, many animals are aware but not strictly conscious. My two German shepherds, for example, are exquisitely aware of and responsive to just about everything around them—more so, in many cases, than I. I *know*, however, that I am conscious because I am aware of my own internal mental state, sometimes even paradoxically aware of that about which I am *un*aware.

On the other hand, I have little doubt that my dogs are conscious, but can't prove it (ditto for my cats and horses). A more satisfying stance, therefore—empathically as well as ethically—is to give in to common sense and stipulate that different animal species possess differing degrees of consciousness. This may be more intellectually satisfying as well, since postulating a continuum of consciousness is consistent with the fundamental evolutionary insight: cross-species, organic continuity.

In any event, the "why" question is as follows: Why should we (or any conscious species) be able to think about our thinking, instead of just plain thinking, period? Why need we know that we know, instead of just knowing? Isn't it enough to feel, without also feeling good—or bad—about the fact that we are feeling? After all, there are downsides to consciousness. For Ernest Becker, "The idea is ludicrous, if it is not monstrous. It means to know that one is food for worms. This is the terror: to have emerged from nothing, to have a name, consciousness of self, deep inner feelings, and excruciating inner yearning for life and self-expression—and with all this yet to die." For Dostoyevsky's

Grand Inquisitor, consciousness and its requisite choices comprise a vast source of human pain (one that he obviated by telling people how to think and what to believe).

There are also some practical problems. As a result of excessive "self-consciousness," we are liable to trip over ourselves, whether literally when attempting to perform some physical act best done via the "flow" of unreflective automaticity, or cognitively, because of the infamous, chattering "monkey mind" so anathematized by Eastern traditions and that may require intense meditation or other disciplines to squelch. Even on a strictly biological basis, consciousness seems hard to justify, if only because it evidently requires a large number of neurons, the elaboration and maintenance of which are bound to be energetically expensive. What is the compensating payoff?

One possibility—a biological null hypothesis—is that maybe consciousness hasn't been selected for at all; maybe it is a nonadaptive by-product of having sufficiently large brains. A single molecule of water, for example, isn't wet. Neither are two, or, presumably, a few thousand, or even a million. But with enough of them, we get wetness—not because wetness is adaptively favored over, say, dryness or bumpiness, but simply as an unavoidable physical consequence of piling up enough H_2O molecules. Can consciousness be like that? Accumulate enough neurons—perhaps because they permit its possessor to integrate numerous sensory inputs and generate complex, variable behavior—wire them up, and presto, they're conscious?

Alternatively, maybe consciousness really *is* adaptive. This would require those who are conscious to be more fit than those lacking this trait; more precisely, genes that contribute to consciousness must somehow have been more successful than alternative alleles in getting themselves projected into the future. Brief explanatory excursion: It is a useful exercise to ask what brains are for. In evolutionary perspective, brains evolved not simply to give us a more accurate view of the world, or merely to orchestrate our internal organs or coordinate our movements, or even our thoughts. Rather, brains exist because they maximize the reproductive success of the genes that helped create them and of the bodies in which they reside. To be adaptive, consciousness must be like that. Insofar as it has evolved via natural selection, consciousness must exist because brains that produced consciousness were evolutionarily favored over those that did not. But why?

One possible avenue of this favoring is that consciousness provided its possessors the capacity to overrule the tyranny of pleasure and pain. Not that pleasure and pain are inherently disadvantageous. Indeed, both have adaptive significance: the former exists as a proximate mechanism encouraging us to engage in activities that are fitness-enhancing, and the latter, to refrain from those that are fitness-reducing. But what about things that are fitness-enhancing in the long run but unavoidably painful in the short? Or vice versa? It might feel good, for example, to overeat, but be nonetheless detrimental. In this case, perhaps a conscious individual can say to himself, "I want to gnaw a bit more on this gazelle leg, but I'd better not." Or vice versa: "I don't want the pain of having my infected tooth pulled out, but if I do it, I'll be better off." Once an individual starts mulling things over, essentially talking to herself about herself, she may be en route to consciousness.

Even more intriguing than consciousness as a facilitator of impulse control, however, is the possibility that it evolved in the context of our social lives, which privileges a kind of Machiavellian intelligence whereby success in competition and cooperation is a function of how well we have evolved the ability to imagine another's situation no less than our own—not so much out of intended benevolence (although this, too, could be the case) as because of the adaptive value of serving one's own evolutionary interests.

Thus, consciousness is not only an unfolding story that we tell ourselves, moment by moment, about what we are doing, feeling, and thinking. It also includes our efforts to interpret what other individuals are doing, feeling, and thinking, as well as how those others are likely to perceive one's self. Call it the Burns benefit, from the last stanza of the Scottish poet's celebrated meditation "To a Louse" . . . : "O wad some Power the giftie gie us/To see oursels as ithers see us!/It wad frae mony a blunder free us/An' foolish notion. . . ."

If, as sometimes suggested, character is what we do when no one is looking, maybe consciousness is precisely a Robert Burnsian evolutionary gift, our anticipation of how we seem to others who *are* looking. And maybe it evolved, accordingly, in the service of our highly developed social intelligence, insofar as this intelligence effectively helped free us from many a blunder and foolish notion, by enabling our consciously endowed ancestors to realize (in proportion as they were conscious) that, for example, seeming too selfish, or insufficiently altruistic, or too cowardly, too uninformed, too ambitious, too sexually voracious, and so forth would ill serve their ends. The more conscious our ancestors were, according to this argument, the more able they were to modify—to their own benefit—others' impressions of them, and hence, their evolutionary success. If so, then genes "for" consciousness would have enjoyed an advantage over alternative genes "for" social obtuseness.

My psychology colleagues have been much exercised of late over "theory of mind," a cognitive mechanism whereby its possessors infer the mental attributes of others. It is a kind of mind reading; not literally, of course, but rather a theory—more accurately, a collection of hypotheses—concerning what is going on inside another's head, so as better to predict his or her behavior. It might be possible to make accurate inferences of this sort without consciousness, but it seems likely that the greater the consciousness by individual A, the more successful she will be in constructing a valid model of the inner workings, and thus the eventual behavior, of individual B. It is one thing to conclude, without reflection "That fellow is angry and hence, dangerous" because of his recent behavior. It is likely to be more fruitful, however, to say—to one's self—something like "He seems angry, just as I was when something similar happened to me. Since I responded in such-and-such a way at that time, I bet he'll respond similarly."

In short, those who possess an accurate theory of mind can model the intentions of others, and profit thereby. And it is at least possible that the more conscious you are, the more accurate is your theory of mind, since cognitive modelers should be more effective if they know, cognitively and self-consciously, not only what they are modeling, but *that* they are doing so.

Just as consciousness doubtless derives at the proximate ("how") level from material events occurring among neurons, the "why" of consciousness is unquestionably a matter of its evolutionary significance, occurring at the level of ecology and natural selection. Nonetheless, many are convinced that consciousness can only have come to us as a gift from God, an endowment enabling His chosen species to glorify the divine and do so with full, aware—that is, conscious—commitment to the saving of their souls. Similarly, there are those who maintain a mystical conception of the power of "cosmic consciousness" to move mountains, or at least, as the Yippies attempted in 1967, to levitate the Pentagon via concentrated psychic energy (an effort that, as I recall, never got off the ground), and/or an unshakable confidence that we are surrounded by disembodied "morphogenetic fields" or other ineffable manifestations of some cerebral happening of which the merely material is only a pale semblance.

"But we are not here concerned with hopes or fears," wrote Darwin, at the end of *The Descent of Man*, "only with the truth as far as our reason allows us to discover it. . . . [W]e must acknowledge, as it seems to me, that man with all his noble qualities, with sympathy which feels for the most debased, with benevolence which extends not only to other men but to the humblest living creature, with his god-like intellect which has penetrated into the movements and constitution of the solar system—with all these exalted powers—Man still bears in his bodily frame the indelible stamp of his lowly origin."

To which I would add that we also bear this stamp—of biology—in our consciousness, not just when it comes to "how" but also "why."

8

Intelligence

CONSCIOUSNESS IS A CONUNDRUM. INTELLIGENCE, BY CONTRAST, should be more straightforward, and by and large, it is. Nonetheless, when it comes to intelligence, some presumably intelligent people hold some very stupid ideas. On one side are those like Charles Murray and the late Richard Herrnstein, whose influential book *The Bell Curve* mangled basic genetic concepts such as heritability, as well as common sense, in their eagerness to conclude that human races differ in IQ. On the other side (politically as well as conceptually), Stephen Jay Gould argued passionately—as well as, on occasion, intelligently—against the legitimacy of intelligence as anything unitary, measurable, or even meaningful.

Whereas "military intelligence" may be an oxymoron, personal intelligence (and even species intelligence) is not. And it is especially ironic for academics to deny the reality of intelligence, since—let's face it—we are downright obsessed with it, whether gossiping about colleagues, evaluating our students, or fuming about George W. Bush. Almost certainly, the truth about intelligence falls between the extremes: It is highly susceptible to environmental influence, and may be complexly multidimensional, thus defying simple assessment, as cogently argued by Howard Gardner, among others. At the same time, intelligence exists as a relatively stable property of individuals, is influenced by genetic factors as well as by environmental ones, emerges from the structure and function of brains, and is consequential for one's life. It also differs significantly among different species, such that what is easy and obvious for one may be totally obscure to another—not because either one is globally smarter than the other, but because its biological needs are different.

And so, when it comes to certain aspects of life in a marsh or swamp, as Kenneth Grahame put it, in *The Wind in the Willows*:

> The clever men at Oxford
> Know all that there is to be knowed.
> But they none of them know one half as much
> As intelligent Mr. Toad

In the bad old days of blinkered thinking, many people—including some evolutionary biologists who should have known better—partook of a misleading dichotomy: genes OR experience, DNA OR culture, instinct OR intelligence. Animals such as toads were supposedly ruled by genes/DNA/instinct, and people, such as Oxford dons, by experience/culture/intelligence. In recent years, the great majority of scientists have come around to the realization that this dichotomy, like so many others (heaven vs hell, organism vs environment, black vs white, and so forth) is a misrepresentation of reality. Things interpenetrate. Every "phenotype"—the observable characteristics of a living thing—derives from the interaction of "genotype" AND environment. When it comes to behavior, this means that every action comes from instinct AND intelligence, acting together. Just as it is meaningless to attribute someone's height to genes or environment, it is also absurd to claim that, say, 3'6" of a woman's stature is due to her ancestry and 2' to her nutrition; every inch of her 5'6" is a consequence of genes AND experience, heredity AND her-experience, acting together.

Moreover, intelligence, which involves the degree to which behavior can be modified by an individual's experience, is itself an adaptive trait, with natural selection having endowed different species with different amounts and kinds. How much, and what kind? Well, that's a matter of which species . . . which is to say, a question of what's adaptive, and for whom. Intestinal parasites are not paragons of intellect, because natural selection hasn't conveyed much advantage to cognitive agility among creatures that live surrounded by food in a dark, homogeneous environment. By contrast, coyotes and ravens and raccoons and human beings are plenty smart—not simply because they have stimulating experiences but because their genes have been endowed, via natural selection, with the ability to build bodies that respond adaptively to such experiences. Intelligence, no less than instinct, is an outgrowth of evolution.

Nonetheless, as to the intelligence of animals—more precisely, the attitudes of people toward the intelligence of animals—we still encounter a peculiarly bimodal distribution of opinion: on the one hand are those who reject it altogether. Intellectual inheritors of René Descartes and Jacques Loeb, they claim that animals are essentially automata, whose behavior can be explained entirely by reflexes or "forced movements," without postulating—or acknowledging—the existence of any internal mental processes. Like Gertrude Stein's famous (and unfair) observation about Oakland, those who deny animals any intellect or subjective mental life maintain, in effect, that "there isn't any there there."

And in the other corner, wearing robes of comparable certainty, the challengers: convinced that pigs postulate, fish philosophize, and, for all one can say, rhododendrons ruminate. Once again, as with the matter of human intelligence, the truth is doubtless in between . . . although several recent studies have been tipping the scales in favor of animal intellect. It's not whether animals have intellect, but which ones, what kind, and how much. And, by implication, whether they are qualitatively different from our own.

Let's start, not surprisingly, with chimpanzees. In 2006, three researchers at the Max Planck Institute for Evolutionary Anthropology, in Leipzig, reported on chimps exposed to two slightly different experimental setups. In one, the apes were able to obtain food by their solo efforts; in the other, they needed the assistance of a second individual. The subject animal could recruit another by literally taking a key, unlocking a door, and releasing the potential assistant. When no collaboration was needed, subjects did not open the door; hence, they obtained the food without being obliged to share. When the coordinated effort of two chimps was required in order to get the food, subjects obtained help. Of special interest—and difficult to interpret without allowing the subjects considerable intellectual acumen—is this finding: When subjects were given the opportunity to choose between two different potential collaborators, kept in adjacent rooms, and when these subjects had been given prior experience in working with each of the two, they preferentially released the one who had previously shown itself to be a more effective collaborator.

There are at least two interconnected parts to the mental agility herein demonstrated. First, recognizing when collaboration is and is not necessary and responding appropriately in each case. Second, ranking the collaborative value of different individuals as a result of previous interactions, remembering their relative value, and also responding differentially and appropriately to those individuals when the circumstance called for doing so.

Nor is impressive intelligence limited to our great ape cousins. Consider the case of Rico, a nine-year-old border collie from Dortmund, Germany, whose mental gymnastics were reported several years previous to the chimp study, once again in the prestigious journal, *Science*, and once again from Leipzig's Max Planck Institute for Evolutionary Anthropology. By the time of the study, Rico had acquired a vocabulary of more than 200 words, most of them nouns, which he demonstrated by successfully retrieving 37 of 40 toys from a collection in a different room, when his name was called by his owner/interlocutor.

In perhaps the most notable test, Rico demonstrated an impressive command of logic. Seven toys, each of whose names Rico knew, were placed in a separate room along with an eighth, which was new to the dog. He was then commanded to return with this eighth item, the name of which he had not previously heard. Seven out of ten times, Rico came back with this eighth, previously unknown, object. Presumably, his internal dialog went something like this: "I'm supposed to fetch the widget, whatever that is. Here are seven things, of which I know that none is a widget. So this one must be a widget." Four weeks later, Rico was able to remember the item three out of six times, comparable to the performance of three-year-old children. Of course, border collies have been bred for close collaboration with human beings, and it remains to be seen whether, for example, Rico can learn *not* to fetch an object upon command.

Nonetheless, "for psychologists," wrote Yale's Paul Bloom in an accompanying article, "dogs may be the new chimpanzees."

Perhaps these intellectual feats are the province of mammals only, with their relatively large brains. This conceit is summarily shattered by the mental exploits of Alex, an African Grey parrot who has been studied for three decades by Irene Pepperberg, currently at Brandeis University. Alex, whose brain is smaller than a walnut, has proven astoundingly adroit, suggesting that birds may be the new dogs.

Recounted in Pepperberg's book, *The Alex Papers*, as well as in numerous filmed appearances, Alex has more than 50 words for different objects. He can also name seven colors (Rico, for all his evident brilliance as a listener/responder, isn't much of a conversationalist), as well as five different shapes, can count to six, and (perhaps most remarkably) he can combine these ideas . . . and they are ideas, not just words for objects, in meaningful ways. Dr. Pepperberg maintains that Alex has the language abilities of a two-year-old child, and the problem solving skills of a six-year-old.

All of which would seem to confirm the suggestion by John Morley, in his *Life of Gladstone*, that "simplicity of character is no hindrance to subtlety of intellect"—except that those people who have been kept by African Grey parrots may well reply that the characters generated by these supposed birdbrains are far from simple!

Given a tray with wooden blocks and wool balls of different colors, and asked "On the tray, how many orange wool?" Alex responds, correctly, "Four." Think about it (Alex evidently did): he must not only discriminate wool from wood, and know which word refers to which, but also distinguish the differently colored wool balls, as well as count them, all in response to one complex request.

When shown an array of things, Alex responds with greater than 80 percent accuracy to questions such as "What object is green and three-corner?" or "What color is wood and four-corner?" or "What shape is paper and purple?" He also understands the concept "different," being able to pick out, upon command, the different one from an array of four things of which three are, for example, of the same color, or three are large and one is small.

Recall the Sesame Street song, "One of These Things (Is Not Like the Others)" designed to hone the reasoning skills of human preschoolers.

Caution is called for when assessing claims of remarkable animal cognitive skills. It is one thing to be generous in interpreting the behavior of other animals, quite another to be taken in. After all, students of animal behavior are still smarting after having been overly credulous about the intellect of Clever Hans, a reputedly brilliant horse. Beginning in the early 1890s, William von Osten astounded European audiences with his horse's ability to answer difficult questions, notably involving arithmetic. Asked, for example, for the sum of three and two, Clever Hans demonstrated his cleverness by tapping his hoof five times. And since it wasn't necessary for von Osten to be present, it seemed most unlikely that any trickery was involved.

But eventually, it was realized that whenever no one in the room knew the answer to a question posed to him, Clever Hans didn't know, either. It turns out that he had learned to respond—by ceasing his hoof tapping—to very subtle nonverbal cues provided by the human beings around him. People would unintentionally tense their muscles until Hans reached the right answer; then they would relax and Hans, sensing this, stopped moving his hoof. Hans was indeed clever, but not as advertised.

These days, researchers in animal behavior are clever, too, and in the studies recounted above—as well as in numerous others accumulating in the scientific literature—they took pains to avoid any hint of unconscious cuing.

Another potential booby trap for researchers in animal intelligence is anthropomorphism, the seductive temptation to attribute human motivation to animals. It is all too easy to describe ants as industrious, lions as lordly, owls as wise. My wife the psychiatrist once suggested—only somewhat in jest—that our African Grey parrot might be depressed, since Oliver (who possesses a large Shakespearean vocabulary) had taken to asking: "To be or not to be," and then concluding, with seemingly mournful mien: "Not to be!" When I recorded Oliver's various soliloquies, however, it was apparent that the constituent phrases were occasionally cut and pasted into patterns that were random and not meaningful in terms of human language or thought.

At the same time, it is now widely understood that sauce for the human goose also works for the animal gander: Intelligence may be mysterious, in the sense of difficult to unravel, but it is no more mystical than any other property of living things. The pioneering geologist

Sir Charles Lyell made an intriguing suggestion when he wrote, in *The Geological Evidence of the Antiquity of Man* (1863), that "so far from having a materialistic tendency, the supposed introduction into the earth at successive geological periods of life—sensation, instinct, the intelligence of the higher mammalia bordering on reason, and lastly, the improvable reason of Man himself—presents us with a picture of the ever-increasing dominion of mind over matter." But notwithstanding his formative influence on Charles Darwin's evolutionary insights, in this instance Lyell was wrong. The question isn't a continuous trajectory of "mind over matter" but rather one of *mind deriving from matter*, whereby the matter of animal brains seems no less capable of generating mind than is the matter of which human brains are composed. The differences are of degree, not kind.

In evolutionary context, intelligence is a biological strategy whereby organisms, human or animal, possess sufficient behavioral complexity and flexibility to respond adaptively to complex situations. Under some conditions, individuals with such qualities were simply more fit than those lacking them, thereby selecting for differing levels of intelligence in certain species, just as natural selection has favored differing patterns of cell metabolism, kidney filtration, or blood circulation.

Also worth noting: There is a small but growing cadre of botanists who argue for the legitimacy of the new field of "plant neurobiology." Plants certainly behave, responding appropriately and integratively to diverse stimuli. In the process, they even make use of electrical and hormonal communication. Whether they are "intelligent" is another matter, but probably not one that should immediately be foreclosed. After all, avocados may be the next parrots.

9

Let Us Reason Together

I TEACH A COURSE AT THE UNIVERSITY OF WASHINGTON TITLED "Ideas of Human Nature." When we talk about reason and rationality, Plato and Aristotle, what Leo Strauss called the world of Athens compared to that of Jerusalem, my students are respectful but restrained; when it's time to deal with unreason and irrationality, Dostoyevsky's Underground Man and some of Stephen Crane's enigmatic poems, they are downright enthusiastic. Was Hamlet therefore wrong? And Aristotle?

"What a piece of work is a man!" exulted the otherwise melancholy Danish prince. "How noble in reason! How infinite in faculty! In form and moving how express and admirable! In action how like an angel! In apprehension how like a god!" Nearly two thousand years earlier, Aristotle maintained that happiness comes from the use of reason, since that is the unique glory and power of humanity. Indeed, for "the Greeks" generally, reason distinguishes us from all other living things, and the life of reason is thus the greatest good to which human beings can aspire. So why doesn't it attract more adherents these days?

For one thing, it may simply be that reason—by definition—is dry and cerebral, only rarely making inroads below the waist. Omar Khayyam made this trade-off uniquely explicit—

> For a new Marriage I did make Carouse:
> Divorced old barren Reason from my Bed,
> And took the Daughter of the Vine to Spouse.

To be sure, excessive reason is easy to caricature, making the daughter of the vine all the more seductive. Thus, at one point in

Jonathan Swift's *Gulliver's Travels*, our hero journeys to Laputa, whose (male) inhabitants are utterly devoted to their intellects: One eye focuses inward and the other upon the stars. Neither looks straight ahead. The Laputans are so cerebral that they cannot hold a normal conversation; their minds wander off into sheer contemplation. They require servants who swat them with special instruments about the mouth and ears, reminding them to speak or listen as needed. Laputans concern themselves only with pure mathematics and equally pure music. Appropriately, they inhabit an island that floats, in ethereal indifference, above the ground. Laputan women, however, are unhappy and regularly cuckold their husbands, who do not notice. The prime minister's wife, for example, repeatedly runs away, preferring to live down on Earth with a drunk who beats her.

Thus presented, to reject reason seems, well, downright reasonable. Consider how rare it is for someone caught in the grip of strong emotion to be overcome by a fit of rationality, but how frequently events go the other way. After all, Blaise Pascal, who abandoned his brilliant study of mathematics to pursue religious contemplation, famously noted: "the heart has its reasons that reason does not understand." Or as 17th-century English churchman and poet Henry Aldrich pointed out in his *Reasons for Drinking*, often we make up our minds first and find "reasons" only later:

> If all be true that I do think,
> There are five reasons we should drink:
> Good wine—a friend—or being dry—
> or lest we should be by and by—
> Or any other reason why!

We may speak admiringly of Greek rationality, of the Age of Reason, and of the Enlightenment, yet it is far easier to find great writing—and even, paradoxically, serious thinking—that extols unreason, irrationality, and the beauty of "following one's heart" rather than one's head. Some of the most "rational" people have done just that.

Legend has it, for example, that when Pythagoras came up with his famous theorem, justly renowned as the cornerstone of geometry (that most logical of mental pursuits), he immediately sacrificed a bull to Apollo. Or think of Isaac Newton: pioneering physicist, both theoretical and empirical, he of the laws of motion and gravity, inventor of calculus, and widely acknowledged as the greatest of all scientists. ("Nature and nature's laws lay hid in Night./ God said, Let Newton be! And all was light.") This same Newton wrote literally thousands of pages, far more than all his physics and mathematics combined, seeking to explicate the prophecies in the Book of Daniel.

Montaigne devoted many of his essays to a skeptical denunciation of the human ability to know anything with certainty. But probably the most influential of reason's opponents was Jean-Jacques Rousseau, who claimed that "the man who thinks is a depraved animal," thereby speaking for what came to be the Romantic movement. But even earlier, many thinkers, including those who employed reason with exquisite precision, had been inclined to put it "in its place." Hardheaded empiricist philosopher David Hume, for example, proclaimed that "reason is, and ought to be, the slave of the passions, and can never pretend to any other office than to serve and obey them." Furthermore, when reason turns against the deeper needs of people, Hume argued, people will turn against reason.

Probably the most articulate, not to mention downright angry, denunciation of human reason is found in the work of Fyodor Dostoyevsky, especially his novella *Notes From Underground*, which depicts a nameless narrator: unattractive, unappealing, and irrational (although intelligent!). In angry contradiction to the utilitarians who argued that society should aim for the "greatest good for the greatest number" and that people can be expected to act in their own best interest, the Underground Man—literature's first "antihero"—jeered that humanity can never be encompassed within a "Crystal Palace" of rationality. He may have a point: Certainly, unreason can be every bit as "human" as the Greeks believed rationality to be. You don't have to be a Freudian, for example, to recognize the importance of the unconscious, which, like an iceberg, not only floats largely below the surface—and is thus inaccessible to rational control—but also constitutes much of our total mental mass.

It is one thing, however, to acknowledge the importance of unreason and irrationality, and quite another to *applaud* it, as the Underground Man does: "I am a sick man. . . . I am a spiteful man. I am a most unpleasant man." The key concept for Dostoyevsky's irrational actor is *spite*, a malicious desire to hurt another without any compensating gain for the perpetrator. Consider the classic formulation of spite: "cutting off your nose to spite your face," disfiguring yourself for "no reason."

Significantly, spiteful behavior does not occur among animals. Even when an animal injures itself or appears to behave irrationally—gnawing off its own paw, killing and eating its offspring—there is typically a biological payoff: freeing oneself from a trap, turning a child (who under certain circumstances may be unlikely to survive) into calories for the parent. Spite is uniquely human.

The Underground Man goes on to rail against a world in which—to his great annoyance—two times two equals four. He claims, instead, that there is pleasure to be found in a toothache, and refers, with something approaching admiration, to Cleopatra's alleged fondness for sticking golden pins in her slave-girls' breasts in order to "take pleasure in their screams and writhing." As the Underground Man sees it, the essence of human-ness is living "according to our own stupid will . . . because it preserves for us what's most important and precious, that is, our personality and our individuality." He believes that people act irrationally because they stubbornly *want* to, snarling that "if you say that one can also calculate all this according to a table, this chaos and darkness, these curses, so that the mere possibility of calculating it all in advance would stop everything and that reason alone would prevail—in that case man would be insane deliberately in order not to have reason, but to have his own way!"

Such sentiments are in no way limited to this most famous apostle of the dark Russian soul or to European Romantics. Here is a poem from that quintessentially American writer, Stephen Crane:

> In the desert
> I saw a creature, naked, bestial,
> Who, squatting upon the ground,
> Held his heart in his hands,
> And ate of it.
> I said, "Is it good, friend?"
> "It is bitter—bitter," he answered,
> "But I like it

Because it is bitter,
And because it is my heart."

But no matter how fashionable it may be to "dis" reason, let's not be carried away. (By what? Presumably, by unreason, since as already suggested, people aren't generally swept away in an uncontrollable fit of rationality.) Strong emotion can be wonderful, especially when it involves love. But it can also be horrible, as when it calls forth hatred, fear, or violence. In any event, one doesn't have to idolize Greek-style rationality to recognize that excesses of unreason typically have little to recommend themselves, and much misery to answer for.

We may admire—albeit surreptitiously—the Underground Man's insistence on being unpredictable, even unpleasant, spiteful, or willfully irrational. But most of us wouldn't choose him to be our financial, vocational, or romantic adviser, or, indeed, any sort of purveyor of wisdom. Maybe unalloyed reason doesn't make the heart sing, but as a guide to action, it is probably a lot better than its darker, danker, likely more destructive, albeit sexier, alternative.

In Newton's case, as in Pythagoras', the most exquisite rationality did not preclude unreason . . . or, as some would prefer to call it, faith. But at least, no great harm seems to have been done by the cohabitation. Sadly, this isn't always the case. "Only part of us is sane," wrote Rebecca West.

> Only part of us loves pleasure and the longer day of happiness, wants to live to our nineties and die in peace, in a house that we built, that shall shelter those who come after us. The other half of us is nearly mad. It prefers the disagreeable to the agreeable, loves pain and its darker night of despair, and wants to die in a catastrophe that will set life back to its beginnings and leave nothing of our house save its blackened foundations. Our bright natures fight in us with this yeasty darkness, and neither part is commonly quite victorious, for we are divided against ourselves.

It may be significant that Ms. West wrote the above while reminiscing on her time in the Balkans, among inhabitants of what we now identify as the former Yugoslavia, people with a long, terrible history of doing things to each other that many outsiders readily label "insane," or at least, "unreasonable." Her point is deeper however, not merely a meditation on Balkan irrationality, but on everyone's.

Take, for a more pedestrian example, the following:

Imagine that you have decided to see a play and paid the admission price of $10 per ticket. As you enter the theater, you discover that you have lost the ticket. The seat was not marked, and the ticket cannot be recovered. Would you pay $10 for another ticket?

Forty-six percent of experimental subjects answered yes; 54 percent answered no. Then a different question was asked: Imagine that you have decided to see a play where admission is $10 per ticket. As you enter the theater, you discover that you have lost a $10 bill. Would you still pay $10 for a ticket for the play?

The results: This time, a whopping 88 percent answered yes and only 12 percent answered no. In other words, most people say that if they had lost their ticket, they would be unwilling to buy another, but if they had simply lost *the value* of the ticket ($10), an overwhelming majority have no qualms about making the purchase! Why such a huge difference? According to psychologists Daniel Kahneman and Amos Tversky (the former a recent economics Nobelist), it is explicable—not by reason but by the way people organize their mental accounts.

Here is another one: "Would you accept a gamble that offers a 10 percent chance to win $95 and a 90 percent chance to lose $5?"

The great majority of people in the study rejected this proposition as a loser. Yet, a bit later, the same individuals were asked this question: "Would you pay $5 to participate in a lottery that offers a 10 percent chance to win $100 and a 90 percent chance to win nothing?" A large proportion of those who refused the first option accepted the second. But the options offer identical outcomes. As Kahneman and Tversky see it: "Thinking of the $5 as a payment makes the venture more acceptable than thinking of the same amount as a loss." It's all a matter of how the situation is framed, in this case, the extent to which people are "risk averse."

Which brings us to yet another perspective on why *Homo sapiens* aren't always strictly sapient. Let's start by agreeing with Herbert Simon (another psychologist who won a Nobel Prize in economics, by the way) that the mind is simply incapable of solving many of the problems posed by the real world, just because the world is big and the mind is small.

But add this: The human mind did not develop as a calculator designed to solve logical problems. Rather, it evolved for a very limited purpose, one not fundamentally different from that of the heart, lungs, or kidneys; that is, the job of the brain is simply to enhance the reproductive success of the body within which it resides. (And in the process, to promote the success of the genes that produced the body: brain and all.)

This is the biological purpose of every mind, human as well as animal, and moreover, it is its *only* purpose. The purpose of the heart is to pump blood, of the lungs to exchange oxygen and carbon dioxide, while the kidneys' work is the elimination of toxic chemicals. The brain's purpose is to direct our internal organs and our external behavior in a way that maximizes our evolutionary success. That's it. Given this, it is remarkable that the human mind is good at solving any problems whatsoever, beyond "Who should I mate with?", "What is that guy up to?", "How can I help my kid?", or "Where are the antelopes hanging out at this time of year?" There is nothing in the biological specifications for brain-building that calls for a device capable of high-powered reasoning, or of solving abstract problems, or even of providing an accurate picture of the "outside" world, beyond what is needed to enable its possessors to thrive and reproduce. Put these requirements, together, on the other hand, and it appears that the result turns out to be a pretty good (that is, rational) calculating device.

In short, the evolutionary design features of the human brain may well hold the key to our penchant for logic as well as illogic. Following is a particularly revealing example, known as the Wason Test.

Imagine that you are confronted with four cards. Each has a letter of the alphabet on one side and a number on the other. You are also told this rule: If there is a vowel on one side, there must be an even number on the other. Your job is to determine which (if any) of the cards must be turned over in order to determine whether the rule is being followed. However, you must only turn over those cards that *require* turning over. Let's say that the four cards are as follows:

T 6 E 9

Which ones should you turn over?

Most people realize that they don't have to inspect the other side of card "T." However, a large proportion respond that the "6" should be inspected. They are wrong: The rule says that if one side is a vowel, the other must be an even number, but nothing about whether an even number must be accompanied by a vowel. (The side opposite a "6" could be a vowel or a consonant; either way, the rule is not violated.) Most people also agree that the "E" must be turned over, since if the other side is not an even number, the rule would be violated. But many people do not realize that the "9" must also be inspected: if its flip side is a vowel, then the rule is violated. So, the correct answer to the above Wason Test is that "T" and "6" should not be turned over, but "E" and "9" should be. Fewer than 20 percent of respondents get it right.

Next, consider this puzzle. You are a bartender at a nightclub where the legal drinking age is 21. Your job is to make sure that this rule is followed: People under 21 must not be drinking alcohol. Toward that end, you can ask individuals their age, or check what they are drinking, but you are required not to be any more intrusive than is absolutely necessary. You are confronted with four different situations, as shown below. In which case (if any) should you ask a patron his or her age, or find out what beverage is being consumed?

#1	#2	#3	#4
Drinking Water	Over 21	Drinking Beer	Under 21

Nearly everyone finds this problem easy. You needn't check the age of person #1, the water-drinker. Similarly, there is no reason to examine the beverage of person #2, who is over 21. But obviously, you had better check the age of person #3, who is drinking beer, just as you need to check the beverage of person #4, who is underage. The point is that this problem set—which is nearly always answered correctly—is logically identical to the earlier set, the one that causes considerable head-scratching, not to mention incorrect answers.

Why is the second problem set so easy, and the first so difficult? This question has been researched by evolutionary psychologist Leda Cosmides. Her answer is that the key isn't logic itself—after all, the two problems are logically equivalent—but how they are positioned in a world of social and biological reality. Thus, whereas the first is a matter of pure reason, disconnected from reality, the second plays into issues of truth-telling and the detection of social cheaters. The human mind, Cosmides points out, is not adapted to solve rarified problems of logic, but is quite refined and powerful when it comes to dealing with matters of cheating and deception. In short, our rationality is bounded by what our brains were constructed—that is, evolved—to do.

One of Goya's most famous paintings is titled "The sleep of reason produces monsters." Monsters, however, arise from many sources, and not just when reason is slumbering and our irrational, unconscious selves have free play. Sometimes, in fact, it is reason itself that generates monstrous outcomes. After all, the gas chambers of Auschwitz and the nuclear devastation of Hiroshima and Nagasaki were technical triumphs, involving no small amount of "rationality." And perhaps I need to acknowledge that no matter the extent to which my students' embrace of the Underground Man seems—to me—downright unreasonable, it is also profoundly human.

10

Believing Is Seeing

I KNOW, I KNOW, IT'S SUPPOSED TO BE THE OTHER WAY AROUND: "Seeing is believing." In any event, let's imagine that instead of being visually oriented primates, we were, say, bloodhounds. In that case, the bromide might have been "Smelling is believing." Or if crickets, perhaps "Hearing is believing." For all we know, limpets (whose lives are spent clinging tenaciously to wave-battered rocks) have their own maxim, to the effect that "Touching is believing." My point is that sensory evidence—seeing, smelling, hearing, touching—generally confirms our knowledge, giving us confidence that something is real. Moreover, we presume that anything real will somehow impinge upon our senses, even if it requires the help of technology, in the form of microscopes, telescopes, and so forth.

Just as a dash, or sometimes a great whopping dose, of irrationality is part of being human, another part of our humanity involves limitations not only of our sense organs but also restrictions in the way we employ our senses no less than our minds.

More often than most people realize, for example, we only "see" things when we are prepared to find them. Consider Edgar Allan Poe's story "The Purloined Letter," in which the eponymous letter escapes detection because it is "hidden" on the suspect's desk, whereas the desperate searchers believe that it must have been secreted somewhere: under the wallpaper, in the floorboards, within the furniture. Their preconceived schema didn't allow that the sought-for object might be sitting there in plain sight, so they didn't see it.

As frequent as this experience may be in daily life, it seems even more common among scientists. This is especially ironic because science is, above all pursuits, the one devoted to perceiving and understanding the phenomenal world. Why, then, might scientists be especially vulnerable to such blindness . . . or at least, blinkered-ness? And what are some examples?

Poe's "purloined letter" was an anomaly; it didn't conform to the searchers' expectation. Thomas Kuhn, in *The Structure of Scientific Revolutions*, maintained that scientific anomalies, observations that exist outside the reigning belief system, are typically ignored until a new conceptual framework (which, in Kuhn's own conceptual framework was designated a "paradigm") becomes the norm. At this point, the anomaly is no longer anomalous, and is therefore recognized. "Discovery," according to Kuhn, "commences with the awareness of anomaly, that is, with the recognition that nature has somehow violated the pre-induced expectations that govern normal science."

Physicists Alan Lightman and Owen Gingerich were, to my knowledge, the first to examine the peculiar connection between recognizing anomalies and accounting for them within the scientific community. They coined the term "retrorecognition," whereby phenomena are only given their due *after* they have been accorded a compelling explanation within a new paradigm. For Lightman and Gingerich, anomalies aren't the cause of a Kuhnian paradigm shift; rather, they are the *result* of such a shift, because under the earlier paradigm, they were unlikely to have been seen at all. An admittedly cumbersome locution, "retrorecognition," nonetheless merits more attention than it has received.

Here are some cases of retrorecognition within my own discipline of evolutionary biology. (I encourage readers to come up with their own examples.)

Much of the impetus behind the Kuhnian paradigmatic revolution known as sociobiology came from William D. Hamilton's identification of inclusive fitness theory. This provided a way to interpret animal and human altruism as behaviors amenable to an intellectually satisfying and theoretically consistent phenomenon, whereby genes maximize their success at projecting copies of themselves into the future. In one of my own textbooks, I now blush to acknowledge, I wrote the following: "Evolutionary biologists, beginning with Darwin, have been troubled by the fact that animals often do things that appear to benefit others, often at great cost to themselves." I, like most sociobiologists, then proceeded to demonstrate how the kaleidoscopic array of animal altruism, previously so bothersome to evolutionary theory, has been normalized by Hamilton's insights into the genetics of kinship.

As it happens, Darwin had indeed been troubled by what he saw as the anomaly of altruism, especially among the "eusocial insects," such as certain bees, wasps, and ants, in which workers forgo personal reproduction and work instead for the reproductive success of someone else: the queen. Armed with Hamilton's crashing insight, it has been possible to make sense of such altruism, as well as numerous other cases, albeit less dramatic but no less intriguing: food sharing, alarm calling, cooperative defense, and so forth, all cases in which seemingly anomalous behavior now fits into a powerful theoretical structure. (After all, natural selection is a fundamentally selfish process, so it is anomalous indeed when individuals do something that *reduces* their genetic representation in future generations. Thanks to "inclusive fitness theory," however, we now understand that what appears to be altruism at the level of bodies can actually be selfishness at the genetic level, and thus, not anomalous at all, at least to those initiated into the paradigm of gene selection.)

But wait! When, just recently, I reviewed textbooks in animal behavior and evolutionary biology written before 1965 (Hamilton's paradigm-shifting paper appeared in 1964) I found that, contrary to my own above-cited canonical textbook assertion, biologists were *not* very much troubled by the occurrence of altruistic behavior. Or if they were, they did not trouble to share their distress in print. Lacking any compelling interpretation for it, they didn't really notice altruism at all. There were, of course, some notable exceptions, such as the renowned neotropical ornithologist Alexander Skutch, who observed "helpers at the nest," and attributed their existence to a need to keep populations in check, while most biologists were oblivious . . . both to the phenomenon and to the illogic of Skutch's explanation. (The problem is that even if some individuals altruistically refrain from breeding in order to maintain the resource base, selfish breeders would profit, and thus spread, whereas any liability would be shared equally.) But now we inevitably point out how strange and anomalous altruistic behavior is, thereby setting the stage for us to explain how it all makes good intra-paradigmatic sense after all, since helpers at the nest, for example, typically help relatives, thereby promoting their own genes. Retro-recognition, anyone?

Another example. A decade or so after Hamilton's inclusive fitness insight, mathematically inclined biologists began turning their attention to game theory, and how it helps explain certain animal interactions. The approach has been fruitful, such that the study of "evolutionarily stable strategies" has by now become well established. The basic idea is that game theoretic analyses—in which payoffs to each "player" are determined, in part, by what another player does—can help explain how separate, distinct behaviors can be maintained in the same population at the same time. Previously, the received wisdom among evolutionary biologists had been that insofar as living things are selected to do whatever maximizes their fitness, all individuals should essentially pursue the same, optimal strategy, differing only in the degree to which they succeed. But now we have a theory that makes sense of situations in which different members of the same population persist in following their own distinct strategies. (In one now-classic formulation, aggressive Hawks and pacific Doves coexist, with neither driving the other to extinction.)

And so it has become commonplace to find reports of evolutionarily stable strategies in all sorts of creatures, doing all sorts of things. Some acorn woodpeckers consistently store acorns, others consistently steal what those others store; some male sunfish are big and courtly, others are small and sneaky; some frogs always croak, others never do; some wasps provision their nests, others enter already-provisioned nests; and so forth. Evolutionary stability, duly retrorecognized, has been popping up everywhere, whereas just a few decades ago, it—like altruism—went altogether ignored.

Examples appear especially prominent in medical diagnoses, probably because a disease syndrome is by definition an anomaly, with normalcy being the default condition. And once again, anomalies tend to be ignored unless, and until, they can be retrorecognized within a grander interpretive framework. Psychiatric diagnoses are particularly susceptible, since, unlike a broken leg or inflamed tonsil, they are notoriously difficult to verify and thus easy to overlook. For a recent example, consider Asperger's Syndrome, a form of mild, high-functioning autism. This diagnosis was unavailable to mental health professionals until 1944, and has only been widely identified since the 1990s. Sure enough, pediatricians, psychiatrists, and psychologists frequently diagnose Asperger's these days, to the degree that support groups and various therapy programs have proliferated. Post-Traumatic Stress Disorder (PTSD) is another familiar case with a similar trajectory.

There is no reason to think that Asperger's Syndrome—or PTSD—isn't real. To say that "believing is seeing" is not to disparage the seeing, or the validity of what is seen. Thus, it is quite different from believers "seeing" Jesus in a tortilla, or Holy Scripture in a cloud. The "seeing" that is facilitated by the existence of an explanatory model, by contrast, is simply a matter of recognizing and taking seriously what genuinely exists but had not previously been acknowledged. Almost certainly, before Asperger's and PTSD had been identified, people nonetheless suffered from these syndromes but were perceived as simply peculiar, disquietingly unusual, or even deeply troubled. But because no diagnosis existed, their problems didn't make sense, and so they weren't really *seen*.

At this point, I must confess to a certain discomfort. Not that I doubt my thesis, but rather, I worry that it might be misconstrued as supporting the postmodernist nonsense that claims to cast doubt upon the "truth claims" of science. So let's be clear: There is a huge difference between acknowledging that certain anomalous phenomena (altruism, evolutionary stable strategies, Asperger's Syndrome, PTSD) aren't recognized until after they can be situated within an interpretive framework and claiming that they don't really exist at all, or that any such framework, because "culturally constructed," is merely self-referential and thus illusory.

Here, it might be useful to refer to a contribution from Paul R. Gross and Norman Levitt, in *Higher Superstition*, their powerful deconstruction of deconstructionism. Gross and Levitt distinguish "soft postmodernism" from its "hard" counterpart. The former—which I readily espouse—acknowledges science to be a social and cultural enterprise, as a result of which the questions asked by practicing scientists are unavoidably constrained by such factors as class, gender, the existing power structure (including, but not limited to, what is "fundable"), and, not least, the existing scientific ethos. But this is a far cry from "hard" postmodernism, which takes the position that cultural constructionism extends to the level of the phenomena themselves, so that belief—and accordingly, bias—are literally all that exist. To hear these folks, there are only "texts," or "narratives," and one is as good as another.

I would wager, by contrast, that before William Harvey—who discovered the function of the heart—people nonetheless died of myocardial infarctions, just as they suffered from angina pectoris, cardiac arrhythmias, etc. They simply didn't understand the source of their problems. Now we do. Not entirely, mind you, as evidenced by ongoing debates over dietary cholesterol, the desirability of moderate alcohol consumption, ideal exercise regimes, the efficacy of bypass surgery, and so forth. But even though our perceptions are strongly colored by social circumstance—including the belief systems of scientists themselves—this is hardly reason to disregard or even devalue the accuracy of these perceptions, once our beliefs have helped our eyes to focus.

At the same time, the notion that "believing is seeing" necessarily casts doubt upon the smug certitude of some scientific knowledge. As a friend of mine likes to point out, "We don't know what we don't know," to which should probably be added "We don't know—or even see—what we don't have a satisfying explanation for." Often I worry that the teaching of science is marred by an overzealous desire to impart to students what we *do* know, leaving them with the impression that everything important is already understood and that science is merely a recitation of that knowledge. The reality, of course, is that there is much more that we don't know, which suggests that perhaps we should teach just that: what we *don't* know about biology, chemistry, geology, or physics.

11

Evolutionary Existentialism and the Meaning of Life

HOW TO BE AN EXPERT IN MONGOLIAN METAPHYSICS? EASY: SPEAK only Mongolian to the metaphysicians and only metaphysics to the Mongolians. The reality, however, is that interdisciplinary studies are more often praised than practiced. It is difficult to bring together seemingly incompatible disciplines, not to mention risky to one's reputation in either. Nonetheless, I am going to try.

For several decades, I have been fascinated by two intellectual challenges that seem to be poles apart: evolutionary biology—especially sociobiology—and existentialism. Avid collectors of oxymorons (jumbo shrimp, freezer burn, civil war) might well lick their chops at the prospect of "evolutionary existentialism." But as I hope to show, although evolutionary biology and existential philosophy seem to be strange bedfellows, they are in fact a remarkably compatible couple.

First, let's turn to some of the more prominent *incompatibilities* between existentialism and sociobiology. Existentialism has, as one of its organizing principles, the basic notion that human beings have no "essence." As Simone de Beauvoir wrote, a person is *l'être donc l'être est de n'être pas*: the being whose essence is having no essence. Or as Jean-Paul Sartre famously put it, "existence precedes essence." For the existentialists, there is no Platonic form of the person, no ideal essence of which our corporeal reality (our physical existence) is a pale instantiation. Rather, we define ourselves, give ourselves meaning, establish our essence, only via our existence, by what we do, by how we choose to live our individual lives. All this because we have no essence (no "human nature") independent of the specifics of how we choose to live.

The concept of choice turns out to be especially important here, because for the existentialists, we are free; indeed, in Sartre's paradoxical words, we are "condemned to be free." Lacking any essence other than our own freedom, we are forced to make choices, and in doing so, we define ourselves. In a huge universe, devoid of purpose and uncaring about people, it is our job to give meaning to our lives by the free, conscious, intentional choices we make.

There is a vast difference between this existentialist conception that there is no human essence, and that presented to us by evolution. Thus, at the heart of an evolution-based conception of human nature—or of hippo, halibut, or hickory tree nature—is the idea that living things are a skin-encapsulated concatenation of genes, jousting with other, similar genes (alternative alleles, more precisely) to get ahead. Free, conscious, intentional choices seem out of place for any creature that is merely the physical manifestation of genes preprogrammed to succeed.

Here is a brief selection from the *Letters* of John Keats, written about 200 years ago, long before sociobiology and even before Darwin. Yet it captures much of the essence (if you'll pardon that word) of modern evolutionary thinking:

> I go amongst the fields and catch a glimpse of a stoat or a fieldmouse peeping out of the withered grass. The creature hath a purpose and its eyes are bright with it. I go amongst the buildings of a city and I see a man hurrying along—to what? The creature hath a purpose and its eyes are bright with it.

For evolutionary biologists, too, living things have a purpose, one that is shared by stoats, field mice, and human beings. What is that purpose? It is, quite simply, the projection of their genes into future generations. Or, as Richard Dawkins and others have emphasized, the behavior of living things can be seen as resulting from the efforts of their constituent genes to get themselves projected into the future. Living things are survival vehicles for their potentially immortal genes.

Let me repeat: biologically speaking, this is what they *are* and it is all that they are.

Most existentialists can be expected to disagree.

For the evolutionary biologist, behavior is one way genes go about promoting themselves. (Other ways are by producing a body that is durable, adapted to its ecological situation, capable of various physiological feats such as growth, metabolism, repair, etc.) This is what behavior *is*. And biologically speaking, it is all that it is. It certainly seems that here—contrary to the existentialist position—is a conception of a human essence, one that is exactly coterminous with human DNA. Moreover, our essence—our genotype—seems in this formulation to *precede* our existence, exactly contrary to what Jean-Paul Sartre and other existentialists would have us believe. And moreover, there doesn't appear to be much room for meaningful freedom of choice—so beloved of the existentialists—insofar as we are slaves to the selfish genes that created us, body and mind.

But stoats and field mice and halibuts and hickory trees don't know what they're doing, or why. Human beings do. Or at least they know what they are doing whenever they let down their guard and allow themselves to be pushed and pulled about by the evolutionary whisperings of their DNA.

"Man is a thinking reed, the weakest to be found in nature," wrote the French mathematical genius, religious mystic, and precursor of existentialism, Blaise Pascal. But he is a thinking reed. It is not necessary for the whole of nature to take up arms to crush him: a puff of smoke, a drop of water, is enough to kill him. But even if the universe should crush him, man would still be more noble than that which destroys him, because he knows that he dies and he realizes the advantage which the universe possesses over him. The universe knows nothing of this.

Thanks to biological insights, people are acquiring a new knowledge: what their own genes are up to, what is their evolutionary "purpose." (As we shall see, this need not necessarily be their existential purpose. Indeed, I will argue that an important benefit of evolutionary wisdom is that by giving us the kind of knowledge about the universe that Pascal so admired, sociobiology leaves us free at last to pursue our own, chosen purposes.)

Pascal prefigured existential thought in other respects as well, as when he wrote that "the silence of these infinite spaces frightens me." Such fear is understandable, since the comfortable sense of human specialness that characterized the pre-Copernican world was being replaced in Pascal's day with a vast universe of astronomic distances, no longer centered on *Homo sapiens*. The great, empty spaces of evolutionary time and possibility—as well as the kinship with "lower" life-forms that it demands—have frightened and repelled many observers of evolutionary biology as well. "Descended from monkeys?" the wife of the Bishop of Worcester is reputed to have exclaimed. "My gracious, let us hope it isn't true. But if it is true, let us hope it doesn't become widely known!"

Well, it is true, and it is becoming widely known. But like Pascal and the existentialists following him, our place in the biological universe is, for many, a chilling reality. "When you look into the void," wrote Nietzsche (a more immediate ancestor of existentialism), "the void looks into you."

Also chilling is the focus—characteristic of both existentialism and modern evolutionary biology—on the smallest possible unit of analysis. Danish philosopher and existential pioneer Søren Kierkegaard asked that this only should be written on his gravestone: "The Individual." And in his masterful *Man in the Modern Age*, the existential psychiatrist and philosopher Karl Jaspers dilated upon the struggle of individuals to achieve an authentic life in the face of pressures for mass conformity.

In a parallel track, much of the intellectual impetus of sociobiology has come from abandoning comfortable but outmoded group-level and "good of the species" arguments, and recognizing that natural selection operates most strongly at the smallest level: notably individuals. (Actually, it goes farther yet, focusing when possible on genes instead.) Its individual and gene-centered perspective has given rise to criticism that sociobiology is inherently cynical, promoting a gloomy, egocentric *weltansicht*. The same, of course, has been said of existentialism, whose stereotypical practitioner is the anguished, angst-ridden loner, wearing a black turtleneck and obsessing, Hamletlike, about the meaninglessness of life.

Let's grant (if only for argument) that human beings—like other living things—are merely survival machines for their genes, lumbering robots whose biologically mandated purpose is neither more nor less than the promulgation of those genes. If so, then there is no more inherent meaning to life as seen in evolutionary terms than when viewed by the existentialists. For most biologists, the promulgation of genes is neither good nor bad. It just *is*. Although scholars (and some scoundrels) have occasionally attempted to derive ethical guidelines from evolution, so-called "evolutionary ethics" has not fared well. This is because such formulations run afoul of what the philosopher G. E. Moore has labeled the "naturalistic fallacy," first elaborated by David Hume: the erroneous expectation that "is implies ought." Although it is tempting to conclude that the natural world provides a model of how human beings ought to behave, it does no such thing. This is most clearly seen when we examine such perfectly natural entities as the virus that causes AIDS: What could be more "organic" than this, made of protein and nucleic acids? Yet, with the exception of those fundamentalist nutcases such as Pat Robertson who claim that AIDS is god's righteous response to sexual sin, no one would characterize HIV as "good."

Those deluded by the promise of evolutionary ethics are probably confusing a laudable bias toward things that are "natural" and "organic" (organic foods, natural childbirth, etc.) with the sense that *anything* natural or organic must be inherently desirable. Most evolutionary biologists, by contrast, know that the natural world—although fascinating—is neither pleasant nor moral. Once again, it just *is*. Indeed, some biologists, notably the eminent theoretician George C. Williams, have emphasized that if we must judge evolution in ethical terms, then if anything, it is downright bad: cruel, selfish, shortsighted, indifferent to the suffering of others, etc. The fruits of evolution, just as the process itself, may inspire our admiration for its complexity and subtlety, but not for any saintliness or even benevolence. Certainly not a "role model." (More on this later.)

It is well known that existentialists are very much occupied with the meaninglessness of life, and the consequent need for people to assert their own meaning, to define themselves against an absurd universe that dictates that ultimately everything will come to naught, because they will die. Less well known—but, I believe, equally valid—is the fact that whereas evolutionary biology makes no claim that it or its productions are inherently good, it—like existentialism—does teach that life is truly absurd.

Evolutionists might well look at all living things—human beings not least—as playing a vast existential roulette game. No one can ever beat the house. There is no option to cash in one's chips and walk away a winner. The only goal is to keep on playing, and indeed, some genes and phyletic lineages manage to stay in the game longer than others. Where, I ask you, is the meaning in a game whose goal is simply to keep on playing, that can never be won, and only lost? And in which we did not even get to write the rules?

In short, there is no intrinsic, evolutionary meaning to being alive. We simply are. And so are our genes. Indeed, we *are* because of our genes, which *are* because their antecedents have avoided being eliminated. We have simply been, as Heidegger (a 20th-century existential precursor of Sartre in particular) put it, "thrown into the world." None of us, after all, was consulted beforehand. Biologically, our genes did it; or rather, our parents' genes. And their parents' before them. Biologists and existentialists might well join in chorusing: How absurd! How meaningless to have been produced in this way, and for such an autistic, self-gratifying purpose; namely, the perpetuation of the genes themselves. Indeed, it isn't really much of a purpose at all, especially because it was never consciously chosen.

Some might say at this point that if evolutionary biology reveals that life is without intrinsic meaning, then this simply demonstrates that biology is mistaken. Not at all. From the perspective of natural science generally, there is no inherent reason why anything—a rock, a waterfall, a halibut, or a human being—is of itself meaningful. Certainly, this is what the existentialists have long emphasized, pointing out that the key to life's meaning is not aliveness itself, but what we attach to it. Kierkegaard, for example, felt passionately about the need for people to *make* their lives meaningful.

The sense of self-construction is far from accidental, because for existentialists there is no reason to suppose that meaning comes pre-packaged along with life itself, even human life. Thus, Kierkegaard once wrote about a man who was so abstract and absentminded that he didn't even realize he existed . . . until one day he woke up and found that he was dead!

To repeat: for evolutionary biology, as for existentialism, there is no inherent purpose or meaning to life. Moreover, a sociobiologic worldview not only denies purpose or meaning, but even ethical guidelines. Insofar as sociobiology helps us identify a kind of goal-directedness—the maximization of genetic representation in the future—it is hardly something that most sentient people are likely to accord the status of "good." It is even something against which sentient human beings can, and, I would argue, *should*, rebel. Indeed, in an already overcrowded planet bursting with 6 billion people, it seems that rebellion is called for.

There are, in fact, many ways that human beings can and do say NO to their genes, just as Sartre and Camus, for example (and Kierkegaard and Nietzsche before them) encouraged people to rebel in their personal lives. We may elect intentional childlessness. We may choose to be less selfish and more genuinely altruistic than our genes might like. We may decide to groom our sons to be nurses and our daughters to be corporate executives. I would even go farther and suggest that we *must* do these sorts of things if we want to be fully human. The alternative—to let biology carry us where it will—is to forgo the responsibility of being human and to be as helpless and abandoned as that (briefly) airborne Magrathean whale we considered some time ago.

"Going with the flow" of our biologically generated inclinations is very close to what Sartre has called "bad faith," wherein people pretend—to themselves and others—that they are *not* free, whereas in fact, they are. Note: This is not to claim that human beings are perfectly free. When philosopher Ortega y Gasset observed, for instance, that "man has no nature, only a history," he neglected that this includes an *evolutionary* history, as the result of which we are constrained as well as impelled in certain ways and directions. We cannot assume the lifestyle of honeybees, for example, or Portuguese men-of-war. But such restrictions are trivial and beside the point, which is that within a remarkable range, our evolutionary bequeathal is almost wildly permissive.

It is interesting, by the way, to consider how much time and energy people expend trying to induce others, especially the young and impressionable, to practice what is widely seen as the cardinal virtue: obedience. To recast Freud's argument about incest restraints: If we were naturally obedient, we probably wouldn't need so much urging to do what we are told. And yet, on balance, it seems that far more harm has been done throughout human history by obedience than by disobedience. I would like to suggest the heretical and admittedly paradoxical notion—based on "evolutionary existentialism"—that, in fact, we need to teach more *dis*obedience. Not only disobedience to political and social authority, but especially disobedience to some of our troublesome genetic inclinations.

The fact that human beings are influenced in various ways by their genes—the subject matter of sociobiology—is, I believe, terribly important, and well worth our study and understanding. As I have already suggested, maybe such understanding is even necessary in order for us to explore the potentials of our own freedom. If this seems incongruous, bear in mind that genetic *influence* is a far cry from genetic *control*. There is very little in the human behavioral repertoire that is under genetic control, very little that is not under genetic influence. At the same time, human beings are remarkably adroit at overcoming such influences.

Consider the game of volleyball. Talented volleyball players do some extraordinary things, making amazing leaps and spectacular saves to keep the ball from touching the ground. In their game can be seen a metaphor for much of human life. The evolutionary imperatives of projecting our genes into the future—technically, maximizing our inclusive fitness—is like the action of gravity on a volleyball. It works, persistently—even remorselessly—in a certain direction. Without substantial efforts on our part, if we stop diving and leaping and batting the ball into the air, gravity wins. But people are incredibly adroit at keeping it airborne. We have invented all sorts of cultural rules, social mores, systems of learning and tradition, some of which support biology and many of which contravene it.

We may not literally define our essence by our existence, as the mid-20th-century existentialists proclaimed, but a deep understanding of sociobiology suggests that the existentialists were absolutely right: Our genes whisper within us; they do not shout. They make suggestions; they do not issue orders. It is our job, our responsibility, to choose whether to obey. We are free, terrifyingly free, to make these decisions, to keep the ball in the air.

Volleyball is a particularly useful metaphor since it also implies a team effort, and human beings are notably social creatures. Sartre indeed observed that "hell is other people," since they interfere with one's freedom, but it remains true that human beings generally do not—and cannot—avoid other human beings. Whether dilemma or delight, our social relationships are also very much the stuff of exciting and important evolutionary insights, illuminating our cooperation with kin and reciprocators, as well as our competition with others.

At the same time, the volleyball image may be troublesome—although no less accurate for that—insofar as it also implies competition: After all, the reason the ball is kept off the ground is so that it can be smashed successfully onto the opponent's court! If desired, feel free to replace a volleyball game with a juggler, working in solitary splendor to keep many different balls in the air.

Time now to introduce another metaphor, this one from existentialism: the Myth of Sisyphus, developed by Albert Camus. You may recall that Sisyphus, a figure from Greek mythology, had irritated the gods and was punished by having to spend eternity pushing a heavy rock up a steep hill, only to have it roll back down again. Sisyphus' punishment, therefore, lasted forever, because his task was never completed. When Camus retold this story in a famous essay, he emphasized that Sisyphus possessed a kind of nobility, precisely because he knew that his efforts were in vain. Sisyphus will never succeed, just as we will never succeed in living forever, in winning the poker game, in contravening our biology. And yet, Sisyphus perseveres. Sisyphus has identified his task, his job in life, knowing its hopelessness and its absurdity. He struggles on anyhow, because that is what it means to be a fulfilled human being. Camus ends his essay by going even further, with the stunning announcement that Sisyphus is *happy*.

I would like to suggest that there are some similarities between the volleyball of human biology, desperately kept above the ground by our various cultural stratagems, and the rock of Sisyphus. In the end, the game is hopeless; biology wins (each of us eventually dies) and the ball bounces, just as the rock rolls downhill. Moreover, as we have seen, the game itself—like the task of Sisyphus—is absurd.

Under the circumstances, perhaps our purpose, our human responsibility, is to make our lives meaningful by emulating Sisyphus. And perhaps here is yet another way that evolution comes in: Just as for the existentialist Camus, Sisyphus achieves a kind of grandeur because he struggles on, fully aware that for him, success is literally impossible—that is, knowing with certainty that the world is stacked against him—I am not alone in suggesting that evolution offers us a kind of potential grandeur as well. Here is the final paragraph of *On the Origin of Species:*

> It is interesting to contemplate a tangled bank, clothed with many plants of many kinds, with birds singing on the bushes, with various insects flitting about, and with worms crawling through the damp earth, and to reflect that these elaborately constructed forms, so different from each other, and dependent upon each other in so complex a manner, have all been produced by laws acting around us. . . . Thus, from the war of nature, from famine and death, the most exalted object which we are capable of conceiving, namely, the production of the higher animals, directly follows. There is grandeur in this view of life . . . that, whilst this planet has gone cycling on according to the fixed law of gravity, from so simple a beginning endless forms most beautiful and most wonderful have been, and are being evolved.

Evolution, especially with its sociobiologic updating, offers us a terrifying insight, a degree of self-knowledge: an understanding of what it is that our genes are up to. Then it leaves us on our own, to decide whether we shall sit back passively, or struggle against this attempted tyranny, like Sisyphus, with all our strength.

Where does this strength come from, and of what does it consist? In one of Plato's dialogues, Socrates comments that we are like marionettes, with the gods pulling our strings at their whim. At the same time, he points out, we have one golden string available by which to pull back, to assert ourselves in return. He was referring to our use of *reason*, by which, according to Socrates, Plato, and generations of Western thinkers ever since, we are to reclaim our identities, our independence, our unique status as autonomous entities.

And yet, if you are fond of ironies, here is a good one: The history of human thought—of the use of this special golden string by which human beings are enabled to distinguish themselves—has led to a progressive debunking of humanity's sense of its own specialness. Intellectual history has been, in a sense, an ongoing series of earthquakes, which add up, in various ways, to a continuing onslaught upon the proposition that human beings are uniquely wonderful. The irony, of course, is that our capacity for complex thought (our ability to pull back against the gods with this golden string) is itself one of our most remarkable—and special—qualities, and yet, at the same time, it is responsible for a *diminution* of our species-wide claim to unique status on Earth, if not in the cosmos.

Thus, with the demolition of the Ptolemaic, earth-centered universe, our planetary home was relegated from centrality to periphery, and *Homo sapiens*, by implication, along with it. But at least *we* remained, self-designated as the apple of God's eye, made in His image. Such a conceit became difficult to maintain, however, with the identification of evolution itself (followed a century later by sociobiology's ongoing elucidation of the behavioral implications of natural selection). Then came Freud's discovery of the unconscious, and the sobering fact that we are not even masters in our own house. Existentialism, too, has been touched by these various intellectual earthquakes, notably Nietzsche's hyper-Darwinian insistence that in a valuesless world, human beings must rise above traditional morality and define themselves as *Übermenschen*. In both a Darwinian and an existential sense, even as our species becomes less central, each individual becomes more.

Recall Plato's golden string, by which we are granted the strength and opportunity to pull back, against the control of the gods. Substitute genes for gods, and you get an oversimplified—indeed, caricatured—evolutionary perspective. Substitute the constraints of society and the inevitability of our death, and you get an oversimplified—indeed, caricatured—existential perspective. Neither discipline has suggested an alternative to Plato's rational rope, although existentialists are especially prone to denigrate the value of "logic chopping," and as we have seen, evolutionary biologists are quick to point out that rationality itself is an adaptation and, as such, situation-specific and often blinkered.

The prospect remains, however, that human beings—despite their biological baggage—retain enough freedom to fashion their own lives and their own future. "The greatest mystery," according to André Malraux, "is not that we have been flung at random among the profusion of the earth and the galaxy of the stars, but that in this prison we can fashion images of ourselves sufficiently powerful to deny our nothingness." Should anyone doubt the capacity of human beings to deny their nothingness and define themselves—if necessary, counter to their evolution-given tendencies—I would like to conclude with a thought experiment that is homey, homely, even scatological, but that should reassure everyone that *Homo sapiens* possesses abundant room for existential freedom.

Begin with this question: Why are human beings so difficult to toilet train, whereas dogs and cats—demonstrably less intelligent than people by virtually all other criteria—are housebroken so easily? Take evolution into account and the answer becomes obvious. Dogs and cats evolved in a two-dimensional world, in which it was highly adaptive for them to avoid fouling their dens. Human beings, as primates, evolved in trees such that the outcome of urination and defecation was not their concern (rather, it was potentially a problem for those poor unfortunates down below!). In fact, modern primates, to this day, are virtually impossible to housebreak.

But does this mean that things are hopeless, that we are the helpless victims of this particular aspect of our human *nature*? Not at all. I would venture that everyone reading this book is toilet-trained! So, despite the fact that it requires going against eons of evolutionary history and a deep-seated primate inclination (or disinclination), human beings are able—given enough training and patience—to act in accord with their enlightened self-interest.

For all its mammalian, evolutionary underpinnings, a primate that can be toilet-trained reveals a dramatic capacity for freedom, maybe even enough to satisfy the most ardent existentialist.

12

The Tyranny of the Natural

IN DISCUSSING CONVERGENCES BETWEEN EVOLUTIONARY BIOLOGY and existentialism, we have gone beyond the question "What is?" to "What should be?" Or, more to the point, "What should we do with our lives?"

It is easy—all too easy—to conclude that if something is natural, it must be good. I, for one, am unabashedly in favor of natural ecosystems and consider them very good indeed. I am also a conscientious consumer of natural foods, a wearer of natural fibers, and a devotee of natural childbirth. And yet, it needs to be said, loud and clear: just because something is natural does not mean it is good. "Smallpox is natural," as Ogden Nash noted. "Vaccine ain't." Ditto for typhoid, tuberculosis, acne, and bladder infections, not to mention hurricanes, tornadoes, tsunamis, and earthquakes: destructive and frequently awful, yet natural as can be, every one.

By the same token, "doing what comes naturally" might be very bad advice indeed.

Nonetheless, people seem to fall (even to jump, enthusiastically) into the misperception identified by David Hume; namely, that "is implies ought." In the early 20th century, philosopher G. E. Moore designated it the "naturalistic fallacy," most eloquently stated, perhaps, by Alexander Pope, in his *Essay On Man*:

> All nature is but art, unknown to thee;
> All chance, direction which thou canst not see;
> All discord, harmony not understood;
> All partial evil, universal good;
> And spite of pride, in erring reason's spite,
> One truth is clear: whatever is is right.

Pope spoke for many. There is a widespread assumption that anything natural is to be applauded and indulged, just as whatever is unnatural must be ethically suspect.

I disagree.

Evolution by natural selection is an extraordinary and endlessly fascinating subject. It has produced you and me and every other living creature. But good it isn't! Physicists to my knowledge have never proposed that the law of gravity, the increase in entropy, or the various electromagnetic "rules" that hold sway among subatomic particles should be consulted as a source of ethical good. If so, we ought to crawl on our bellies, hold fast to anyone different from ourselves—as positive adheres to negative—and never clean our rooms. My first point, accordingly, is that natural selection is every bit as natural as Newton's Laws, Planck's constant, or relativity, and every bit as devoid of moral direction. Like the laws of physics, the laws of biology simply describe what is, not what should be.

My second point is that if anything, the evolutionary process is more negative than neutral when it comes to humane values; it is likely to lead to results that most ethicists will, and should, reject. (Note: I am NOT counseling a rejection of natural selection or of evolution as a historical, biological, and natural process; rather, I reject the intimation that natural selection is somehow a moral exemplar.)

Our current understanding of natural selection is that it operates as a ratio, with the numerator reflecting the success of genes in projecting copies of themselves into the future and the denominator, the success of alternative alleles. Since a gene (or an individual, a population, even—in theory—a species) maximizes its success by producing the largest such ratio, it can do so either by reducing the denominator or increasing the numerator. Most creatures, most of the time, find it easier to do the latter than the former, which is why living things generally are more concerned with feathering their nests than de-feathering those of others.

Taken by itself, such self-regard isn't the stuff to make an ethicist's heart go pitter-patter. But to make matters worse, animal studies in recent years have revealed a vast panoply of behavior whereby living things have no hesitation in minimizing the denominator, trampling over others in pursuit of their own biological benefit. We have long known that the natural world is replete with grisly cases of predation, parasitism, a universe of ghastly horrors all generated by natural selection and unleavened by the slightest ethical qualms on the part of perpetrators.

In her stunning Pulitzer Prize–winning memoir, *Pilgrim at Tinker Creek*, Annie Dillard described her horror at watching a frog whose innards were liquefied and then sucked dry by a giant water bug. Ms. Dillard also shared her puzzled outrage at the phenomenal wastefulness of an evolutionary process that generates hundreds, often thousands, of tiny but perfect lives, only to snuff most of them out, relentlessly and heartlessly.

Worse yet, perhaps, are the cases of vicious genetic self-promotion at the expense of others. For example, biologists have documented infanticide in numerous species, including lions and many nonhuman primates such as langur monkeys and chimpanzees. The basic pattern is that when a dominant male is overthrown, his replacement often systematically kills the nursing infants (unrelated to himself), thereby inducing the lactating mothers to resume their sexual cycling, whereupon they mate with their infants' murderer. It is truly awful, such that even hard-eyed biologists had a difficult time accepting its ubiquity, and even—until recently—its "naturalness." But natural it is, and a readily understood consequence of natural selection as a mindless, automatic, and value-free process, whose driving principle is if anything not just amoral but—by any decent human standard—downright immoral.

Add cases of animal rape, deception, nepotism, siblicide, matricide, and cannibalism, and it should be clear that natural selection has blindly, mechanically, yet effectively favored self-betterment and self-promotion, unmitigated by any ethical considerations. I say this fully aware of an important recent trend in animal behavior research: the demonstration that animals often reconcile, make peace, and cooperate; no less than the morally repulsive examples just cited, these behaviors also reflect the profound self-centeredness of the evolutionary process. If the outcome in certain cases is less reprehensible than outright slaughter, it is only because natural selection only sometimes works to reduce the denominator of the "fitness ratio." Most of the time, it increases the numerator. But all the time, the only outcome assessed by natural selection is whether a given tactic works—whether it enhances fitness—not whether it is good, right, just, admirable or in any sense moral. Why, then, should we look to such a process for moral guidance? Indeed, insofar as evolution has engendered behavioral tendencies within ourselves that are callously indifferent to anything but self- (and gene-) betterment, and armed as we now are with insight into the origin of such tendencies, wouldn't sound moral guidance suggest that we intentionally act contrary to them?

In the movie *The African Queen* (based on an even better book by C. S. Forester), Katharine Hepburn stiffly observes to Humphrey Bogart: "Nature, Mr. Allnutt, is what we are put in this world to rise above." I strongly doubt that we were put on earth to do anything in particular, but if we want to be ethical—rather than simply "successful"—rising above our human nature may be just what is needed. Evolution by natural selection, in short, is a wonderful thing to learn about . . . but a terrible thing to learn from.

By the end of the 19th century, Thomas Huxley was perhaps the most famous living biologist, renowned in the English-speaking world as "Darwin's bulldog" for his fierce and determined defense of natural selection. But he defended evolution as a scientific explanation, NOT as a moral touchstone. In 1893, Huxley made this especially clear in a lecture titled *Evolution and Ethics*, delivered to a packed house at Oxford University. "The practice of that which is ethically best," he stated,

> what we call goodness or virtue—involves a course of conduct which, in all respects, is opposed to that which leads to success in the cosmic struggle for existence. In place of ruthless self-assertion it demands self-restraint; in place of thrusting aside, or treading down, all competitors, it requires that the individual shall not merely respect, but shall help his fellows; its influence is directed, not so much to the survival of the fittest, as to the fitting of as many as possible to survive.

"The ethical progress of society depends," according to Huxley, "not on imitating the cosmic process, [that is, evolution by natural selection] still less in running away from it, but in combating it."

It may seem impossible for human beings to "combat" evolution, since *Homo sapiens*—no less than every other species—is one of its products. And yet, Huxley's exhortation is not unrealistic. It seems likely, for example, that to some extent each of us undergoes a trajectory of decreasing selfishness and increasing altruism as we grow up, beginning with the infantile conviction that the world exists solely for our personal gratification and then, over time, experiencing the mellowing of increased wisdom and perspective as we become aware of the other lives around us, which are not all oriented toward ourselves. In her novel *Middlemarch*, George Eliot noted that "we are all born in moral stupidity, taking the world as an udder with which to feed ourselves." Over time, this "moral stupidity" is replaced—in varying degrees—with ethical acuity, the sharpness of which can largely be judged by the amount of unselfish altruism that is generated.

Many sober, highly intelligent scientists and humanists misunderstand the connection between evolution and morality, grimly determined that evolutionary facts are dangerous because they justify human misbehavior. Developmental psychologist Jerome Kagan exemplifies this blind spot. "Evolutionary arguments," he writes, "are used to cleanse greed, promiscuity, and the abuse of stepchildren of moral taint." Similar arguments were in fact used in this way, in the unlamented days of social Darwinism. But no longer. Professor Kagan is living in the lamentable past.

Today, evolutionary thinking is used to *understand* greed, promiscuity, and the abuse of stepchildren . . . and also to help understand parenting, nepotism, reciprocity, friendship, parent-offspring conflict, courtship, violence, love, adultery, altruism, and bigotry, to name just a few.

Human beings, more than any other living things, are characterized by an almost unlimited repertoire, a behavioral range that exceeds that of any other living creature. It is well within our capacity to say "No" to our evolutionary bequeathal, especially once we recognize its unethical underpinnings. After all, we engage in all sorts of activities that are unnatural, but good. Some of them require going directly against some human inclinations, and, although not easy to achieve, are readily within the human repertoire once the social and personal benefits are made clear and compliance demanded. Others, like playing the violin, learning a second language, or composing a novel, require effort and dedication. In this sense, once again, they are not "natural" like learning to walk or eating when hungry. But they are not only achievable, they can be some of humankind's greatest accomplishments, natural or not. In fact, a case can be made that those human achievements that are greatest, most noteworthy, most lasting and sublime are those that are achieved when people act contrary to their "natural" inclinations. "Drink when you are not thirsty," we are advised in Mozart's opera, *The Marriage of Figaro*. "Make love when you don't want to—this is what distinguishes us from the beasts." I'm not sure about the wisdom of drinking or making love when disinclined, but it's hard to argue with the suggestion that writing an opera—unnatural as it may be—is nonetheless a good idea.

The point is this: Whatever our biological tendencies for selfish, unethical, altogether "natural" behavior, there is reason for optimism about our capacity to rise above such inclinations, especially if we recognize the wisdom of doing so. Not everyone is cut out to write great opera, but the good news is that every human being– as a result of being human—is capable of overcoming the tyranny of the natural. "We Shall Overcome," indeed! In fact, the capability of overcoming may be a reasonable definition of what it *means* to be human.

13

Forbidden Knowledge?

SOCRATES WAS MADE TO DRINK HEMLOCK FOR HAVING "CORRUPTED the youth of Athens." Is sociobiology or—as it is more commonly called these days—"evolutionary psychology," similarly corrupting? Although the study of evolution is one of the most exciting and illuminating of all intellectual enterprises, there is at the same time something dark about the implications of natural selection for our own behavior. Insofar as evolutionary biology serves up some bad news about our own inclinations, ought we to suppress it? If, as argued previous chapter, the "natural" is often nasty, what about the downside of letting such nastiness become widely known?

Should we revise Pink Floyd's anthem *We Don't Need No Education*—with its chorus "No dark sarcasm in the classroom/ Teacher, leave those kids alone"—to "No dark sociobiology in the classroom"? To answer this, we need first to examine that purported darkness.

Basically, it's a matter of selfishness. For a long time, evolution was thought to operate "for the good of the species," a conception that had a number of pro-social implications; this, in turn, may be one reason why "species benefit" was so widely accepted, and why its overthrow took so long and was so vigorously resisted. Thus, if evolution somehow cares about the benefit enjoyed by a species, or of any other group larger than the individual, then it makes sense for natural selection to favor actions that contribute positively to that larger whole, even at the expense of the individual in question. Doing good therefore becomes doubly right: not just ethically correct but also biologically appropriate. In a world motivated by concern for the group rather than the individual, altruism is to be expected, since it would be "only natural" for an individual to suffer costs—and to do so willingly—so long as other species members come out ahead as a result.

Then came the revolution. Beginning in the 1960s with a series of paradigm-shifting papers by William D. Hamilton, a notable book by George C. Williams (*Natural Selection and Adaptation*), and then, further clarifications in the early 1970s, especially by Robert L. Trivers and John Maynard Smith, and magisterially summarized in Edward O. Wilson's *Sociobiology*, the conceptual structure of modern evolutionary biology was changed . . . maybe not forever (it's a bit premature to conclude that), but into the foreseeable future. Sociobiology was born on the wings of this scientific paradigm shift, whose underlying manifesto holds that the evolutionary process works most effectively at the smallest unit: that of individuals and genes, rather than groups and species.

At first glance, none of this seems especially threatening. Moreover, it has been liberating in the extreme, shedding new light on a wide range of animal and human social behavior. But at the same time, the individual- and gene-centered view of life offers, in a sense, a perspective that is profoundly selfish; hence, Richard Dawkins's immensely influential and thought-clarifying book, *The Selfish Gene*. The basic idea has been so productive that it has rapidly become dogma: living things compete with each other (more precisely, their constituent genes struggle with alternative copies) in a never-ending process of differential reproduction, using their bodies as vehicles, or tools, for achieving success. The result has been to validate a view of human motivations that seems to approve personal selfishness while casting doubt on any self-abnegating actions, seeing a self-serving component behind any act, no matter how altruistic it might appear. Sociobiologists have thus become modern-day descendants of the cynical King Gama, from Gilbert and Sullivan's *Princess Ida*, who proudly announces his cynicism: "A charitable action I can skillfully dissect; And interested motives I'm delighted to detect."

Scientifically, such "detection" works. Ethically, however, it stinks: if the fundamental nature of living things—human beings included—is to joust endlessly with each other, each seeking to get ahead, then we're all mired in selfishness. A dark vision indeed.

* * *

It might ease the blow by noting that such a vision of human nature is hardly unique to modern evolutionary science. Thus, in *An Enquiry Concerning Human Understanding* (1748), David Hume wrote that

> should a traveller, returning from a far country, bring us an account of men . . . wholly different from any with whom we were ever acquainted . . . who were entirely divested of avarice, ambition, or revenge; who knew no pleasure but friendship, generosity, and public spirit; we should immediately, from these circumstances, detect the falsehood, and prove him a liar, with the same certainty as if he had stuffed his narration with stories of centaurs and dragons, miracles and prodigies.

Hume also noted, albeit playfully, "It is not irrational for me to prefer the destruction of half the world to the pricking of my finger." More than 200 years ago, people were discomfited by such sentiments, and they still are.

Just as nature is said to abhor a vacuum, it abhors true altruism. Society, on the other hand, adores it. Most ethical systems advocate undiscriminating altruism: "Virtue," we are advised, "is its own reward." Such sentiments are immensely attractive, not only because they are how we would like other people to behave, but probably because at some level we wish that we could do the same. As Bertolt Brecht notes in *The Threepenny Opera*, "We crave to be more kindly than we are," so much so that purveyors of good news—those who proclaim the "better angels of our nature"—nearly always receive a more enthusiastic reception than do those whose message is more dour.

Although people are widely urged to be kind, moral, altruistic, and so forth—which suggests that they are basically less kind, moral, altruistic, etc., than is desired—it is also common to give at least lip service to the precept that people are fundamentally good. It appears that there is a payoff in claiming—if not acting—as though others are good at heart. "Each of us will be well advised, on some suitable occasion," wrote Freud, in *Civilization and Its Discontents*, "to make a low bow to the deeply moral nature of mankind; it will help us to be generally popular and much will be forgiven us for it." Why are people generally so unkind to those who criticize the human species as being, at heart, unkind? Maybe because of worry that such critics might be seeking to justify their own unpleasantness by pointing to a general unpleasantness on the part of others. And maybe also because most people like to think of themselves as benevolent and altruistic, or at least, to think that other people think of them this way. It seems likely that a cynic is harder to bamboozle.

In *Civilization and Its Discontents*, perhaps his most pessimistic book, Freud went on to lament that one of education's sins is that

> it does not prepare them [children] for the aggressiveness of which they are destined to become the objects. In sending the young into life with such a false psychological orientation, education is behaving as though one were to equip people starting on a Polar expedition with summer clothing and maps of the Italian Lakes. In this it becomes evident that a certain misuse is being made of ethical demands. The strictness of those demands would not do so much harm if education were to say: "This is how men ought to be, in order to be happy and to make others happy; but you have to reckon on their not being like that." Instead of this the young are made to believe that everyone else fulfills those ethical demands—that is, that everyone else is virtuous. It is on this that the demand is based that the young, too, shall become virtuous.

At the same time, we can expect that society will often call for real altruism, not because it is good for the altruist but because it benefits those who receive. If it were clearly good for the altruist, then society wouldn't have to call for it! In fact, cynics are prone to pointing out that it is precisely because altruism is generally not good for the altruist that social pressures are so often focused on producing it. Friedrich Nietzsche was probably the most articulate spokesperson for the view that society encourages self-sacrifice because the unselfish sucker is an asset to others:

> virtues (such as industriousness, obedience, chastity, piety, justness) are mostly injurious to their possessors. . . . If you possess a virtue . . . you are its victim! But that is precisely why your neighbor praises your virtue. Praise of the selfless, sacrificing, virtuous . . . is in any event not a product of the spirit of selflessness! One's 'neighbor' praises selflessness *because he derives advantage from it*." [italics in original]

If Nietzsche is correct, then there is a distressingly manipulative quality to morals, to most religious teachings, to the newspaper headlines that celebrate the hero who leaps into a raging river to rescue a drowning child, to local Good Citizenship Awards and PTA prizes.

"That man is good who does good to others," wrote the 17th-century French moralist Jean de la Bruyère. Nothing objectionable so far; indeed, it makes sense (especially for the "others"). But de la Bruyère goes on, revealing a wicked pre-Nietzschean cynicism:

> If he suffers on account of the good he does, he is very good; if he suffers at the hands of those to whom he has done good, then his goodness is so great that it could be enhanced only by greater suffering; and if he should die at their hands, his virtue can go no further; it is heroic, it is perfect.

Such "perfect" heroism can only be wished on one's worst enemies.

Exhortations to extreme selflessness are easy to parody, as not only unrealistic but also paradoxically self-serving insofar as the exhorter is likely to benefit at the expense of the one exhorted. Yet, the more we learn about biology, the more sensible becomes the basic thrust of social ethics, precisely because nearly everyone, left to his or her devices, is likely to be selfish, probably more than is good for the rest of us. Philosopher and mathematician Bertrand Russell pointed out that "by the cultivation of large and generous desires . . . men can be brought to act more than they do at present in a manner that is consistent with the general happiness of mankind." Society is therefore left with the responsibility to do a lot of cultivating.

Seen this way, a biologically appropriate wisdom begins to emerge from the various commandments and moral injunctions, nearly all of which can at least be interpreted as trying to get people to behave "better," that is, to develop and then act upon large and generous desires, to strive to be more amiable, more altruistic, less competitive, and less selfish than they might otherwise be.

Enter sociobiology. With its increasingly clear demonstration that Hume, Freud, Brecht, Nietzsche (also Machiavelli and Hobbes) are basically onto something, and that selfishness resides in our very genes, it would seem not only that evolution is a dispiriting guide to human behavior, but also that the teaching of sociobiology (or evolutionary psychology) should only be undertaken with great caution. Renowned primatologist Sarah Hrdy accordingly questioned "whether sociobiology should be taught at the high school level . . . because it can be very threatening to students still in the process of shaping their own priorities," adding that "the whole message of sociobiology is oriented toward the success of the individual. . . . [U]nless a student has a moral framework already in place, we could be producing social monsters by teaching this."

What to do?

One possibility—unacceptable, I would hope, to most people—would be to refrain altogether from teaching such "dangerous truths." Teacher, leave those kids alone! Preferable, I submit, is to structure the teaching of sociobiology along the lines of sex ed: teach what we know, but do so in age-appropriate stages. Just as we would not bombard kindergartners with the details of condom use, we probably ought not instruct pre-teens in the finer points of sociobiology, especially since many of these are hidden even to those expected to do the teaching. For one thing, a deeper grasp of the evolutionary biology of altruism reveals that even though selfishness may well underlie much of our behavior, it is often achieved, paradoxically, via acts of altruism, as when individuals behave in a manner that enhances the ultimate success of genetic relatives: Here, selfishness at the level of genes produces altruism at the level of bodies.

Ditto for "reciprocity," which, as Robert Trivers elegantly demonstrated more than three decades ago, can produce seemingly altruistic exchanges and moral obligations even between nonrelatives. Yet, genetic selfishness underlies it all. Alexander Pope concluded, with some satisfaction: "That REASON, PASSION, answer one great aim; That true SELF-LOVE and SOCIAL are the same."

Sociobiologists understand that there is an altruistic as well as a selfish side to the evolutionary coin. A half-baked introduction to the discipline, which pointed only to the latter, would therefore do students a substantial disservice. Moreover, gene-centered evolutionary thinking can also expand the sense of self and emphasize interrelatedness: Altruism aside, just consider all those genes for cellular metabolism, for neurotransmitters and basic body plans, all of them shared with every living thing, competing and pushing and somehow working things out on a small and increasingly crowded planet. There, by the grace of evolution, goes a large part of "ourselves."

"Gene-centered theories are often reviled," writes gene theorist David Haig,

> because of their perceived implications for human societies. But, even though genes may cajole, deceive, cheat, swindle or steal, all in pursuit of their own replication, this does not mean that people must be similarly self-interested. Organisms are collective entities (like firms, communes, unions, charities, teams) and the behaviours and decisions of collective bodies need not mirror those of their individual members.

To some extent, in short, we may even possess—gulp!—Free Will.

Beyond the question of what our genes may be up to and the extent to which we are independent of them, those expected to ponder the biology of their own "natural" inclinations ought also to be warned (more than once) about the "naturalistic fallacy," the presumption that things natural are, ipso facto, good. I'd even suggest pushing this farther, and say that the real test of our humanity might be whether we are willing, at least on occasion, to say "No" to our "natural" inclinations, thereby refusing go along with our selfish genes. To my knowledge, no other animal species is capable of doing this. More than any living things, we are characterized by an almost unlimited repertoire; human beings are of the wilderness, with beasts inside, but much of this beastliness involves gene-based altruism no less than selfishness. (Recall the paradox that genetic selfishness is often promoted via altruism toward other individuals insofar as these recipients are likely to carry identical copies of the genes in question.)

Moreover, as Carl Sandburg put it, each human being is "the keeper of his zoo." Even this is not evidence of a lack of evolutionary influence; rather, it is a result of selection for being a good zookeeper. Socrates, we are told, elected to drink the hemlock when he could have followed a different path. Human beings are capable not only of understanding the route along which evolution has placed them, but also of deciding, in the clear light of reason as well as ethics, whether to follow this path.

14

Are We Selfish Altruists? Group-Oriented Individualists? (Or What?)

A HUGE OCTOPUS EMERGES FROM THE OCEAN, WRAPS AN OVERSIZED tentacle around the waist of a young woman, and proceeds to drag her into the sea. This memorable episode from Thomas Pynchon's vast, surreal novel, *Gravity's Rainbow*, has a happy ending, however, owing to the intervention of Mr. Tyrone Slothrop, who first unavailingly beats the molluscan monster over the head with an empty wine bottle. Then, in a stroke of zoologically informed genius, our hero offers the briny behemoth something even more alluring than a fair maiden: a delicious crab. It works, suggesting that this particular octopus conforms, at least in its dietary preference, to the norm for its species. Nonetheless, we learn that "in their brief time together, Slothrop formed the impression that this octopus was not in good mental health."

It isn't at all clear where the creature's mental derangement lies. The octopus in question actually behaved with a reasonable degree of healthy, enlightened self-interest in seeking first to consume the young lady, and then forgoing her for the even more delectable crab. Nonetheless, nature writer David Quammen may have been onto something when he pointed out that octopi generally—and not just Thomas Pynchon's admittedly fictional creation—might be especially vulnerable to mental disequilibrium, if only because one of their distinguishing characteristics is having immense brains. Mental strain is probably not unknown among animals, but there seems little doubt that it is especially well developed in the species *Homo sapiens*, whose brains are especially large, and whose strain is correspondingly great.

The cause of our curious cerebral hypertrophy has been the subject of much speculation and research, including hypotheses that it facilitates communication, tool use, dealing with predators as well as prey, even the prospect that our braininess might be a byproduct of sexual selection, comparable to the peacock's flamboyant tail feathers. I'd like to suggest yet another possibility: perhaps our massive mentation evolved because of the peculiar pressures of keeping a very complex social life in adaptive equilibrium. This possibly harebrained schema for explaining our human-brained selves has at least one virtue: it also speaks to a long-standing question in ethics, one that might also be illuminated—at least in part—by evolutionary biology. That question is how to navigate the conflicting demands of personal desire versus social obligation.

Once again, and not for the last time, we find ourselves confronting the question of altruism versus selfishness, one of the signature themes of modern evolutionary biology. We shall stick with it for the next several chapters (the issue will stick with you, and me, and everyone else, long after this book is forgotten).

As difficult as it must be for any creature to balance its various competing requirements—to eat or sleep, to attack or retreat, etc.—such demands are probably greatest in the domain of social life. As confusing and stressful as it must be to predict the vagaries of weather, for example, the vagaries of one's fellow creatures have to be even more complex, confusing, and stressful. And when it comes to having a complicated and difficult social life, human beings are in a class by themselves. Clearly, our remarkably oversized brains do not satisfy themselves with simply meeting the contingencies of daily life. Human neurons are obsessed with confronting all sorts of difficult issues, mostly of their own making. Small wonder that so many people, like Pynchon's octopus, are not in good mental health.

Some of the most anguished dilemmas human beings encounter derive from conflicting loyalties between what we want to do and what we feel that we should do, between crosscutting obligations toward ourselves on the one hand and our family on the other, toward one family member versus another, toward friends versus the larger community, toward the law, the nation, other living things, the planet.

Unlike Thomas Pynchon's giant octopus, we need something more than a juicy and distracting morsel to avoid inflicting pain on others. We need to refrain both from morally repulsive excesses of selfishness, as well as from overdoses of self-destructive altruism. One of the most powerful insights of current evolutionary theory is that our brains have been produced by self-serving genes, yet these same self-serving inclinations have resulted in behavior that is often cooperative, social, and—at the level of bodies if not genes—altruistic. At the same time, we confront ourselves in societies that are, at best, uneasy compromises among the competing selfish tendencies of its component parts. The conflicting pressures of selfishness and altruism are so difficult to unravel that perhaps our species can only restore its mental health by employing that wonderful brain to reflect upon its own evolutionary situation. After all, *Homo sapiens* includes not only its nasty "animal" aspects of violence and selfishness but also those positive, equally animal components of benevolence and even self-sacrificing altruism.

This stubborn contention between selfishness and social obligation generates what we might call "Maggie's Dilemma," after the heroine of George Eliot's novel, *The Mill on the Floss*. Maggie could become the wife of either of two attractive young men (the selfish, personally fulfilling route), but at the cost of mortifying her family, especially her rigid and disapproving brother. Or she could deny both prospective husbands (and, thus, herself), while remaining true to her social obligations. Maggie's Dilemma is stated by Eliot as follows: "The great problem of the shifting relation between passion [selfishness] and duty [social altruism] is clear to no man. . . ." Maggie resolves it in favor of the latter: "I cannot take a good for myself that has been wrung out of their [her family's] misery."

For most of us, Maggie's Dilemma remains very real. Gratify yourself, or your family? Cheat a little here and there or be an upstanding, honest person? Discard your trash or recycle it? Be bad and satisfy your "passion" or be good, and do your "duty"?

Whether because of the normal unfolding of our "innate" altruism or the gradual success of such ethical exhortations, most people come evenually to realize that they aren't the center of the universe, and that their needs and and desires aren't necessarily paramount. It is widely assumed that this transition, from selfishness to increasing altruism, is universal. Thus, although people are widely urged to be kind, moral, altruistic, and so forth—which suggests that they are basically *less* kind, moral, altruistic, etc., than is desired—it is also common to give at least lip service to the precept that people are fundamentally good.

On the other hand, there may be some wisdom in the naïvely optimistic miseducation that Freud found so troublesome, if only because its alternative—the expectation that others will be aggressive and nasty—can become a self-fulfilling prophecy, especially if it leads people to be aggressive and nasty first.

The human paradox is even more complicated. In *The Ghost in the Machine*, Arthur Koestler suggested that violence is not caused so much by an excess of individualistic selfishness as by too much group-centeredness; in a sense, an excess of altruism:

> The total identification of the individual with the group . . . makes him perform comradely, altruistic, heroic actions—to the point of self-sacrifice—and at the same time behave with ruthless cruelty towards the enemy or victim of the group. But the brutality displayed by the members of a fanatic crowd is impersonal and unselfish; it is exercised in the interest or the supposed interest of the whole; and it entails the readiness not only to kill but also to die in its name.

By this point, it should be clear that in generating human beings, evolution has put together a strange amalgam of selfish nastiness and altruistic kindliness, sometimes fading into group-oriented, altruistic violence. Neither altruism nor selfishness is more "biological" than the other, and yet, the argument that human beings are naturally group-oriented, cooperative, or altruistic is widely seen as less "instinctivist" or "genetically determinist" than the alternative view: that we are all somewhat competitive, aggressive, and selfish. Biology does not have a monopoly on nasty, selfish behavior, nor does social learning work only on behalf of altruism. People can learn violence and they are just as much "naturally" altruistic (especially toward kin and reciprocators) as they are "naturally" selfish.

One of the oldest debates among philosophers, ethicists, and theologians concerns this fundamental division of human nature: Are people naturally generous, altruistic, group-oriented, pro-social; that is, basically good? Or are we nasty, selfish, always looking out for private gain; that is, basically bad? Interestingly, evolution doesn't answer this question so much as italicize it. Thus, insofar as a genetic perspective is accurate, genes are in fact selfish, if only because their biological function is to promote their own replication. But the paradoxical take-home lesson of sociobiology is that selfishness is often achieved by an array of altruistic, pro-social acts: toward relatives (who, by definition, have a certain probability of carrying the "altruistic" genes in question), reciprocators (past or potential), even, on occasion, the larger group. It is a matter of seeing the glass of human selfishness as either half full or half empty.

In short, whereas the genes that make up every human being are fundamentally selfish, expressing this selfishness can result in a kind of group-oriented altruism. Kin selection, for example, produces a powerful bias toward family members, including, perhaps, others who are psychologically identified as kin, even though they aren't. The urge for reciprocity generates another powerful current, flowing toward the exchange of favors and moral obligations (i.e., friendship). Reproduction and parental behavior lead to remarkable levels of self-sacrifice and devotion. But, genetic selfishness underlies it all.

And yet, even behavioral tendencies that are generally regarded as moral and desirable can fade imperceptibly into actions that are immoral and undesirable: Excessive concern with one's evolutionary success—to the detriment of others who aren't related—results in complaints of nepotism, something generally seen as unattractive, unfair, often illegal. Too much pro-social identification with the group can breed not only patriotism, but chauvinism, jingoism, bigotry, and warmongering. Reciprocity can lead to cheating, and parenting, to manipulation and conflict.

No one said these issues would be easy. As noted earlier, perhaps this is one reason why human beings often find their mental health under assault. In one *Sesame Street* song, Kermit the Frog points out "It's not easy bein' green." It's not easy being human, either.

Take the remarkable opening pages of the novel *Enduring Love* by Ian McEwan. Joe is enjoying a picnic in the British countryside when he hears a shout for help and discovers a man struggling with a large gas balloon, being tossed about by the wind. There is a little boy in the basket. Joe and four other men grab the balloon by a trailing rope, but, just when it seems that they are going to rescue the boy, a sudden powerful gust of wind carries the balloon and its occupant over the edge of an impossibly steep slope. Joe and three of the other men let go immediately; the fourth holds on, but not for long. He falls to his death, having tried to save the boy (who, ironically, manages to survive uninjured).

As Joe reflects on the event—and how he and the three other men had released the rope, choosing to save themselves rather than the child—he acknowledges its primordial quality:

> This is our mammalian conflict, what to give to others and what to keep for yourself. Treading that line, keeping the others in check and being kept in check by them, is what we call morality. Hanging a few feet above the Chilterns escarpment, our crew enacted morality's ancient dilemma: us, or me.

In this case, Joe didn't know the boy in the balloon, and certainly wasn't related to him. Therefore, "us" didn't outweigh "me." On the other hand, the same was true for the man who died, yet he didn't let go . . . until it was too late. Maybe he was following a different, "higher" morality (lethally elevated, a cynic might add). Or maybe he just held on too long, then couldn't let go safely even if—when?—he wanted to.

Part of the difficulty of being human is the often agonizing need to decide on one's own ethical precepts, to establish the boundaries of what is desirable, the limits of what is acceptable, and when, and why. Can sociobiology—a gene's-eye view of evolution—help? Hard to say. Maybe it will assist, if only in helping to clarify Maggie's Dilemma (how to navigate between the Scylla of selfish desire and the

Charybdis of social obligation), and thus, assisting human beings in their unending quest to Know Themselves. Or it might hurt, if people take selfishness as its primary lesson (and especially if they go further and commit the "naturalistic fallacy" and assume that anything that is "natural" is necessarily good). On the other hand, if people focus instead on the altruistic side of the evolutionary coin—on the pro-social, reciprocating, kin- and group-oriented aspect of human nature—they are likely to derive a very different take-home message: There, by the grace of evolution, go a large part of "ourselves," part hungry octopus, part Tyrone Slothrop (also, part edible crab), part altruistic saint and selfish sinner, by turns bighearted and narrow-minded.

15

Dealing with Dilemmas: Personal Gain versus Public Good

FEW ISSUES ARE AS VEXED—IN PUBLIC POLITICS, PERSONAL ETHICS, and, not coincidentally, evolutionary science—as the matter of altruism versus selfishness or of individual self-interest versus the greater good. The resulting dilemmas are immediate and practical, as well as conceptual.

I teach a seminar in which I restrict enrollment to fifteen students. Others typically want to get in, and yet, ironically, much of the class's popularity is due to the benefits that come from keeping it small. Most students understand the advantages of small classes of this sort, and so they wouldn't want everyone who wishes admittance to get in; just themselves! The larger the enrollment, of course, the more is discussion inhibited, to everyone's disadvantage. And so, each year I find myself in the difficult position of telling a number of students that there simply isn't room for them. Each student turned away from this class understands the logic, but nonetheless, each would like the limit to be expanded . . . by just one.

This is what experts in game theory call a social dilemma: a "prisoner's dilemma" writ large, in which individuals "play" against the larger collectivity. It has a lot to teach us.

On the one hand, the class as a whole is somewhat worse off for every student above a given number (arbitrarily set in this case at 15) who is admitted. On the other, each student who wants admission would be better off—or expects to be—as the 16th. Individuals seeking admission are playing against the rest of the class, and most are willing to "defect," in game theory terms, by getting in, all the while assuming that the instructor will not allow *everyone* who wants entry to succeed. In fact, if I allowed unrestricted access to the class, it would become a large lecture, losing its special value to all concerned.

In such cases, the downside for the group is generally distributed across many individuals, so the personal cost to the defector (in our example, someone who successfully "overloads" into the class) is likely to be low, whereas the benefit, for this person alone, is likely to be high enough that, on balance, he or she is best off being selfish. The dilemma is that if everyone is selfish, then all are worse off, although each individual is tempted to try to get away with it nonetheless.

Let's take a big leap in scale, to global warming. Instead of "add yourself to David Barash's class," make it "add extra carbon dioxide to the earth's atmosphere." The accumulation of greenhouse gases in the atmosphere is having negative effects on the world's climate; this is obvious to all but industry lobbyists, a very small number of contrarian (and/or financially compromised) scientists, and free market–worshipping ideologues. Technical solutions to global warming exist; the real dilemma is social. It is often easier, and, in the short run, cheaper, to use polluting, carbon dioxide–spewing devices than to refrain—even though if everyone does it, we're all worse off. So, at the individual level, most people would prefer to drive their private automobiles rather than take public transportation, all the while bemoaning the resulting traffic, not to mention the ensuing buildup of carbon dioxide and its effects.

At the corporate level, firms are reluctant to cut back on their generation of greenhouse gases because it may place them at a competitive disadvantage. The United States under George W. Bush refused to abide by the Kyoto Accords, complaining that it was not in the "national interest," even though clearly in the interest of the planet. The United States, in this case, elected to defect, "playing against" most other countries. In such situations, the cost to each entity—person, corporation, country—of cooperating appears to be high, whereas the benefit seems low. Defection threatens to become the rule, whereupon everyone loses.

This is precisely what happens in many cases.

Consider a water shortage. Individuals should cooperate and

conserve, but each is inclined to cheat. After all, many people like a lush lawn, soapy showers, frequent flushes. People readily understand that the entire public—which includes themselves—would be in trouble if everyone else indulged his or her private desires. But it is awfully tempting to cheat, because the cost of each personal defection, to the group as a whole, is small, whereas the benefit to the cheating individual is potentially large, so long as only a few others do the same thing.

A similar situation applies to taxes. No one likes paying them. At the same time, nearly everyone recognizes the benefits of living in a society in which people pay up: It is beyond the capacities of individual citizens to maintain highways, libraries, fire departments, police forces, national defense, schools, hospitals, and so forth. And yet, it is awfully tempting to try cheating the IRS, or certainly, to err on the side of personal benefit, even though if no one paid his or her share, we'd all be in big trouble. Getting away with an unfairly small tax bill may constitute a major personal windfall, and in a country with predictable tax revenues in the trillions of dollars, the temptation to defect is powerful indeed.

Defection happens. For example, in 2000, Stephen King announced that he would make his next book available online. *The New York Times* reported that "Mr. King is trusting his readers to send him a dollar after each download. . . . If he does not receive payments for at least 75 percent of the downloads, he says, he will stop writing after two chapters, and readers won't learn the end." It happened that Mr. King's electronic novel, *The Plant*, generated more than 120,000 downloads when the first chapter appeared; by the mid-November installment, only 40,000 were being downloaded, and of these, only 46 percent were paid for. As a result, Stephen King stopped writing it. "If you pay, the story rolls. If you don't, the story folds," King had written on his Web site. It folded.

In fact, once you start identifying social dilemmas, it's difficult not to see them everywhere, whether in public affairs or private matters. Some more examples: ordering an expensive meal when it has been agreed that a group will divide the bill evenly; hoarding the drug Cipro immediately after the 2001 anthrax scare, when supplies were short; refusing to join a trade union and thus avoiding dues, but profiting from whatever benefits the union obtains for members and nonmembers alike; rushing for an exit during a theater fire (everyone's chances are better if each individual files out in an orderly manner, but you would probably get out more quickly if you pushed others aside); not giving to National Public Radio or your local blood bank; or even doing something as trivial as standing on tiptoe during a parade (you'll get a better view, but others will be obstructed).

Nearly forty years ago, ecologist Garrett Hardin wrote a now-classic article titled "The Tragedy of the Commons," in which he pointed to the history of land abuse in England. In this case, land held in common was overgrazed by livestock owners who realized that by doing so, the commons was diminished—to everyone's detriment—but who felt, nevertheless, tempted to graze their own animals there, out of fear that if they refrained, others would take advantage of the resource, and it would be ruined anyhow.

During the period of communist-style governments in the Soviet Union and Eastern Europe, the environment was no better protected there than in the capitalist West; indeed, by most measures, it was worse. Socialist production goals were accorded the highest priority and, as a result, environmental values suffered greatly. International meetings of environmentalists were notable for the hopes of the participants, often pinned unrealistically on the opposing system. Thus, Western environmentalists expected that state control and authority would generate models of more reliable environmental protection, while environmentalists from the Soviet bloc had an equally idealistic—and unrequited—hope that private ownership might lead the way toward rational ecological stewardship.

It has been said that under capitalism, people exploit people, whereas under communism, it's the other way around! Either way, the environment has been dangerously abused, and to a large extent, social dilemmas are to blame.

Comparable temptations are evidently felt by animals, too. Reproduction, in a strictly Darwinian perspective, is selfish. After all, baby-making is the primary way living things project their genes into the future, thereby receiving an evolutionary payoff . . . actually, *the* evolutionary payoff. Consider, for example, African elephants: Increasingly restricted to game parks, they are often fenced in, and, as a result, locally overcrowded. Desperate to satisfy their huge appetites, hungry elephants strip the bark from trees, eventually killing them and thus, ultimately destroying their habitat, to their own long-term detriment. But try explaining that to the elephants. For eons, natural selection has rewarded those who—selfishly and successfully—reproduced. Once again, it's a kind of social dilemma, whereby every elephant who becomes a parent gains a payoff measured in evolutionary fitness, while in the process imposing a substantial long-term cost on their habitat and thus, on elephants as a whole.

Biologists were intrigued when Robert Trivers, at the time a Harvard University graduate student, pointed out that "reciprocal altruism" can be expected among animals. But despite more than three decades of serious effort to document self-sacrificing behavior among nonrelatives, remarkably few such examples have been discovered. We must conclude that the temptation to defect is not uniquely human. Birds do it. Bees do it. Even monkeys in their trees do it. They give in to temptation and typically refrain from giving to the group, or at least, from giving more than they have to. In fact, it may be that if anything, animals are *less* generous—more inclined to be selfish—than their human counterparts, because they don't have powerful ethical precepts (religion, morality, not to mention the IRS and the criminal courts) to remind them of their social obligations.

Mallard ducks, those lovely, iridescent green-headed denizens of freshwater ponds, provide an especially chilling example. Drake mallards are notorious rapists, which is to say, they force copulations with already mated ducks whose participation is obviously not consensual. These sexual attacks frequently involve many males, and are the mallard equivalent of gang rape among human beings. In the process, females suffer a high mortality, because their heads are held under water during copulation, and in the course of multiple, sequential rapes, they can drown. Why, then, do male number two, and three, and so forth persist in forcing their sexual attentions on a female who may well die as a result? Clearly, this is a bad payoff for these males, and yet, each attacking drake is stuck in a social dilemma: to refrain, after other males have gone ahead, would be to increase the chance that the victimized female will survive, but with the guarantee that she will not be conceiving the "cooperator's" darling little ducklings. And so, male mallards "defect" and participate in gang rapes of females, to the detriment of the victimized females and even, to some extent, of themselves, stuck in a social dilemma of their own making, but one that is nonetheless devastatingly real.

Social dilemmas are unavoidable. After all, individuals only rarely exist in isolation. Nearly always, we interact with the rest of society, expected to cooperate, yet tempted to cheat, relying on the cooperation of others, yet vulnerable to their defection. The basic concept of "society" assumes give and take, a "social contract" whereby individuals make what is essentially a deal with society at large: Each will forgo certain selfish, personal opportunities in exchange for profiting from the cooperation of others.

Theories of social contract in relationship to social dilemmas have occupied many of the great thinkers in political philosophy. The question, in short, is simply this: How to reconcile personal selfishness with public benefit?

Thomas Hobbes lobbied forcefully for the necessity of regulating selfish, nasty human impulses for the good of the larger whole. Although social dilemmas had not been identified as such in Hobbes's time, he clearly saw that the seductive power of social defection was dangerously strong. In *Leviathan*, Hobbes wrote that "during the time men live without a common Power to keep them all in awe, they are in that condition which is called Warre; and such a warre, as if of every man, against every man." The role of the political sovereign, in game theory terms, was to punish noncooperators and keep everyone in line, preventing each from defecting. Hobbes envisioned that without such control, we would all inhabit a state of nature in which people were incapable of arranging for such cooperative endeavors as agriculture, industry, arts, or even society itself, "and which is worst of all, continual feare, and danger of violent death; And the life of man, solitary, poore, nasty, brutish and short."

With his idealization of the "noble savage," Jean-Jacques Rousseau seems poles apart from Hobbes. Yet Rousseau, too, in his best-known work, *The Social Contract*, pointed in a similar direction. Rousseau made an important distinction, between the "will of all" (the sum of individual desires) and the "general will" (the good of society, taken as a whole), emphasizing that the social contract is a way of making sure that pursuit of the former doesn't destroy the latter. Rousseau begins his famous book noting that "man is born free, and is everywhere in chains." Although he often inveighs against these chains, even Rousseau recognizes that they are necessary:

> In order that the social contract should not be a vain formula, it tacitly includes an undertaking, which alone can give force to the others, that whoever refuses to obey the general will shall be constrained by the whole body. . . . The undertakings that bind us to the social body are obligatory only because they are mutual.

In short, the peculiar genius of society is that it forces people to abide by their social contracts, and allows them to bypass the siren call of social defection, by precisely the kind of restraints and restrictions that Hobbes recommended and that we might expect Rousseau—given his adulation of the "noble savage," untrammeled by rules and regulations, conventions, and compulsions—to have opposed. But even Rousseau, apostle of natural human inclinations, recognized the need to say NO to the temptations posed by our many social dilemmas. Two centuries later, in "The Tragedy of the Commons," Garrett Hardin found himself arguing similarly, coming down in favor of "mutual coercion mutually agreed upon."

There is, however, a problem; namely, the mandating of restraints and enforcement mechanisms to prevent selfish, socially irresponsible behavior (the recommendation of Hobbes, Rousseau, and Hardin), runs counter to what is probably *the* basic precept of free market capitalism. Thus, in the most famous paragraph in his masterpiece, *The Wealth of Nations*, Adam Smith introduced the notion of the "invisible hand," whereby the common good is achieved most efficiently when each individual succumbs to private greed:

> It is not from the benevolence of the butcher, the brewer, or the baker, that we expect our dinner, but from their regard to their own self-interest. We address ourselves, not to their humanity but to their self-love, and never talk to them of our own necessities but of their advantages. . . . It is his own advantage, indeed, and not that of society which he has in view. . . . He intends only his own gain, and he is

> in this, as in many other cases, led by an invisible hand to promote an end which was no part of his intention. . . . By pursuing his own interest he frequently promotes that of the society more effectually than when he really intends to promote it.

Political conservatives love this sort of stuff, which gives them permission to extol personal selfishness with the claim that by pursuing private gain individuals are also promoting the public good. Game theory in general and social dilemmas in particular help point out that this is self-serving nonsense.

Faced with a choice between private gain and public good, most people opt for the former, and yet, such a response to social dilemmas—itself consistent with conservative political philosophy insofar as it emphasizes the "fallen" aspect of human nature—suggests that, if anything, the all-too-visible hand of personal defection is likely to result in disaster rather than benefit to society as a whole. I am not going to suggest a way out because as far as I am concerned, the solution is apparent, and available, even as it is abhorrent to free market fundamentalists: the urgings, requirements, and restraints of society as a whole; that is, the benevolent intervention of government, demanding at least a modicum of cooperation on behalf of society and the greater good.

After all, we accept any number of impositions on our personal freedom: It may be in my selfish interest to rob a bank, if I could get away with it. But it isn't in society's interest for there to be lots of bank robbers, so we agree that police, courts, and jails are necessary, to make it disadvantageous for individuals to defect in this particular social dilemma. As we have seen, Hobbes—generally considered a political conservative, incidentally—saw this situation as requiring that individuals hobble their freedom in their own, collective, long-term interest, just as I find myself required to act as benevolent doorman to students seeking admission to my seminar. Doesn't the same argument apply to the worst excesses of unfettered greed in the public arena?

When it comes to altruism and selfishness, it turns out that very little is truly "unfettered." Just as John Muir once noted that "when we try to pick out anything by itself, we find it hitched to everything else in the universe," when we pick out altruism, we find it hitched to selfishness—and even something worse.

16

The Ugly Underside of Altruism

THE WORLD'S FIRST TELEGRAPHIC MESSAGE, SENT BY SAMUEL F. B. Morse on May 24, 1844, was a question: "What hath God wrought?" Thinking now of human beings, many are asking, "What hath evolution wrought?" And the answer, not surprisingly, is complex and contradictory, appropriate for a creature among whom—as we have seen—social and self-love, altruism and selfishness, niceness and nastiness, bigheartedness and nastiness interpenetrate, like yin and yang.

This answer comes to us because of a revolution in evolution, or, rather, in biologists' understanding of how natural selection works. This revolution derives from a new sense of what is important, what is the biological "bottom line," what is—as evolutionists put it—the appropriate "unit of selection." More and more, it has become clear that the crucial unit is the gene: not the species, nor the group, nor even the individual, but the smallest meaningful entity that can persist through evolutionary time. After all, genes are potentially immortal, whereas individuals come and go. As Richard Dawkins so brilliantly emphasized in his book *The Selfish Gene*, living things are essentially constructed by their genes, for their (the genes') benefit. And when individuals behave, such actions are "adaptive" insofar as they contribute, ultimately, to the success of those genes in promoting copies of themselves into the future. (Equally important: Activities that do not contribute to genetic success are selected against.)

Accordingly, biologists had been perplexed on occasion to find animals behaving altruistically, doing things that helped others to survive and reproduce, but at some cost to the altruist, such as giving an alarm call when a predator approached (thereby aiding the listeners but at some cost to the alarmgivers, who are rendered more conspicuous), or sharing food, or simply tolerating a freeloader.

The problem for altruism, and thus, for evolutionary biologists, was simple: Evolution rewards selfishness. Insofar as a trait or behavior increases reproductive success, that trait or behavior should become more abundant, along with its corresponding gene(s). At the same time, any trait or behavior or gene(s) that reduced reproductive success should quickly disappear, to be replaced with its selfish alternatives.

Natural selection, in short, helps those who help themselves. And it penalizes those who help others. As a result, biologists were troubled—not ethically, mind you, but as scientists—by the very fact of altruism's perseverance, since it should quickly be selected against and replaced by selfishness, which, by definition, helps itself and thus prospers. Another way of stating the problem: How to explain the endurance of traits that are, by definition, self-defeating?

But endure they do. It turns out that altruism not only astounds, it abounds. Why?

Here is where the revolution comes in. Part of the charm of the gene's-eye perspective is that it solved much of the altruism question. Thanks to English biologist William D. Hamilton, the paradox of altruism was resolved by revealing that it wasn't really a paradox after all! Hamilton's crashing insight was that if individual altruists direct their benevolence preferentially toward others who are close relatives, then genes are actually benefiting themselves, and that this process is, literally, what natural selection is all about. Hamilton thus pointed out that what often appears to be altruism at the level of bodies can actually be selfishness at the level of genes, which benefit themselves by proxy.

"How do I love thee? Let me count thy genes."

Isaac Newton opened physicists' eyes to why things fall; Hamilton, in effect, opened biologists' eyes to why living things behave as they do, even what they are. Hamilton's now-classic article "The Genetical Evolution of Social Behavior," published in 1964, is, more than any other single piece of research, the intellectual cornerstone of the modern evolutionary revolution. In effect, Hamilton's insight was to recognize that genes promote their success via copies of themselves in other bodies.

Even before Hamilton, biologists had never been troubled, interestingly, by the ubiquity of reproduction, even though *at the level of bodies*, breeding is just as altruistic as alarm-calling or food-sharing. After all, reproduction is costly. It takes time and energy. It involves risk and imposes penalties on the would-be breeder. Think of the time and energy spent in courtship, the vulnerability associated with mating, the sheer metabolic cost of constructing a placenta, lactating, defending and provisioning one's offspring, etc. Reproducing, in short, benefits someone else—the offspring—while it imposes a cost on the parent.

Yet parental behavior is not normally considered altruistic; making children is not surprising, nor is it in any way counterintuitive, or against what an evolutionary biologist—or anyone else—might expect. Quite the opposite: Most people take reproduction for granted, and biologists have long considered that successful breeding is central to evolutionary success. For decades, in fact, biologists equated breeding with "fitness." Reproduction is costly? Of course. But it would be absurd to think that as a result, reproduction would be selected against! What would replace it? All living things are the offspring of parents who successfully reproduced, costs and all. A genetic basis for nonreproduction would have a dim evolutionary future indeed.

But here's the point: At the gene level, the important thing about reproduction is that genes are packaging copies of themselves into new bodies, and then (in varying ways, depending on the species) trying to promote the success of these new bodies. How? By feeding them, keeping them warm, protecting them, teaching them, taking them to soccer games and the orthodontist, maybe sending them to college and paying their bills. With this new perspective, having babies and then caring for them is seen for what it is: a perfectly good route to evolutionary success. But not the only route. Hamilton's genius came in recognizing that there are other ways for genes to be successful, if not via those bodies that we call offspring, then via other bodies that we call genetic relatives, such as nieces, nephews, cousins, grandchildren, etc.

The only difference between these more distant relatives and those that we call offspring is that the more distant the relative, the lower the probability that a gene present in any given individual is also present in that relative. This, after all, is what people mean when they talk about a "distant" relative: genetic distance, even though most of us lack the ability to calculate exactly how great the distance and precisely what genetic "distance" really means. (No matter: most of us don't understand the details of neurophysiology, either, yet are pretty good at thinking.) Hamilton went further, showing the conditions necessary for altruism to evolve.

To make a long mathematical story short, "Hamilton's rule" is that altruism will be selected for in proportion as (1) the cost to the altruist is low; (2) the benefit to the recipient is high; and (3) the altruist and recipient are closely related. The first condition means that low-cost altruism—for example, taking a small risk for someone else—should be easier, and thus more frequent, than running extreme risks. The second means that all things being equal, altruists should be more likely to act in proportion as their altruism helps the individual being assisted. And the third condition means that "more closely related = more altruism" and conversely, "less closely related = less altruism." Why? Because the closer the genetic relationship, the higher the probability that any altruism-promoting gene(s) present in the altruist will also be present in the recipient.

The result is a new picture of evolutionary fitness, one that reveals how the net of natural selection is spread more widely than pre-Hamiltonian Darwinists had imagined. Previously, when biologists thought about fitness, they considered only direct reproductive success; its importance has never been in doubt. But breeding is only part of the story. The full tale, known as "inclusive fitness," is more, well, inclusive. It includes not only reproductive success but also any action that increases another's survival and reproduction. But—and here is a very important point—not all "others" are equal, at least insofar as a would-be altruist is concerned. The importance of each "other" (to the altruist) is greater in proportion as he or she is more closely related, because that means a greater probability of shared genes.

Actually, there are two noteworthy precursors of Hamilton's important insight. Both were brilliant evolutionists—in fact, they were two of the most prominent founders of the field of population genetics—but for some reason, neither carried this particular idea very far. In the late 1920s, Ronald A. Fisher wondered why certain bad-tasting caterpillars were brightly colored. He acknowledged that conspicuous coloration would make it more likely that a hungry bird, for example, after eating one caterpillar, would leave the others alone. But, Fisher pointed out, such an "advantage" would come a bit late for the caterpillar who sacrificed its life in order to educate predators not to make the same mistake twice. Fisher went further, suggesting that perhaps this is why such insects tend to be found in groups: If these groups consist of brothers and sisters, then the dying caterpillar (rather, the relevant genes within the caterpillar) would be repaid—not in this life, but in evolutionary time—through the success of kin.

The other biologist who caught a glimpse of the genetics of altruism but apparently did not realize its generalizability was J. B. S. Haldane, like his contemporary Fisher and his successor Hamilton, a British mathematics whiz. The story goes that Haldane was at his favorite pub when the conversation happened upon self-sacrificial bravery. Haldane was asked if he would give his life for his brother. No, he said, he wouldn't do that. Then he made a rapid calculation on the back of a napkin and added that he'd do so for two brothers or eight cousins! (Genes within any of us enjoy a 1/2 probability of occurring within a full sibling; hence, two brothers equals one self. Similarly, cousins are, on average, 1/8 genetically identical, so it takes eight cousins to comprise the genetic equivalent of one's self.)

This, apparently, is as far as the realization went, until Hamilton revisited the paradox of altruism, bequeathing us a new view of ourselves and of life more generally. The result is also sometimes called "kin selection," since it speaks to a predictable bias toward kin: relatives over nonrelatives, and closer relatives over more distant ones. Kin selection—or "inclusive fitness theory"—suggests that nepotism is likely to be universal, or nearly so, in the living world. It even provides a way of calculating it. Thus, one self equals two brothers, or four grandchildren, or eight cousins, etc. Faced with the question, "Save your skin or save your kin?", the balance point occurs when the likelihood of genes present in relatives equals that within one's self.

Armed with this new view of behavior, biologists began reinterpreting the living world.

And by and large, predictions based on Hamilton's "inclusive fitness" model have been confirmed. Across a remarkable range of species and a wide array of behaviors, animals preferentially direct beneficence toward relatives over nonrelatives, also favoring close relatives over distant relations. Not only that, but thinking in terms of shared genes has helped elucidate such "cross-cultural universals" as nepotism among human beings. It even refocuses basic understanding of life itself, shedding new light, for example, on why multicellular bodies remain as coherent as they do. (After all, why should the liver cells uncomplainingly undertake the unpleasant task of detoxifying the blood, leaving all the evolutionary success to the gonads? Because liver and gonad cells are genetically identical, so that success for the latter leads to exactly the same triumph for all other body cells.)

There are certainly additional factors that underpin altruism, in human beings as well as other animals. Thus, reciprocity is sometimes important, and occasionally what appears to be altruism is simply selfishness—even at the personal level. Nonetheless, Hamilton's insight into the significance of shared genes and altruism has been so powerful that it can fairly be identified as one of the greatest advances in modern evolutionary theory.

I assume that most readers are with me at this point, even though some may part company at the assertion that insights derived from animals apply—albeit with reservations—to human beings. But what of altruism's "ugly underside"?

After all, a vision of Darwinian competition without shared genes to soften the blows is far more unpleasant. In his book, *The Economy of Nature and the Evolution of Sex*, marine biologist and historian of science Michael Ghiselin put it chillingly and well:

> No hint of genuine charity ameliorates our vision of society, once sentimentalism has been laid aside. What passes for cooperation turns out to be a mixture of opportunism and exploitation. . . . Where it is in his own interest, every organism may reasonably be expected to aid his fellows. Where he has no alternative, he submits to the yoke of communal servitude. Yet given a full chance to act in his own interest, nothing but expediency will restrain him from brutalizing, from maiming, from murdering—his brother, his mate, his parent, or his child. Scratch an "altruist," and watch a "hypocrite" bleed.

By contrast, gene-based altruism seems downright delightful.

Of course, to some people, nepotism is itself ugly. That's why we have laws against carrying it too far. And to others, it is demeaning to consider that something as lofty as altruism may have an underlying selfish component. Reductionism may be the stuff of science (at least, most science), but when applied to understanding ourselves, it often fails to make the heart sing. What's really unpleasant about the biology of gene-centered altruism, however, is much more troublesome, and—if true—far more deserving of universal condemnation.

It is this. Insofar as shared genes underpin much of human altruism, the apparent absence of shared genes may well lead to altruism's nasty inverted doppelgänger: intolerance, hatred, and bigotry. "How do I hate thee? Let me count thy genes."

If genes are predisposed to behave nicely toward identical copies of themselves housed in other bodies, they presumably have ways of achieving this identification. For some species, simple physical proximity may do the trick: Close neighbors are somewhat more likely to be relatives. For others, behavioral cues may be available: Someone in your nest, or den, or household is likely to be more closely related to you than is someone in a different social unit. (Sociologists have long been intrigued by "in-group amity, out-group enmity"; now biologists are, too.) There also remains the possibility that genes predispose their bodies to behave benevolently toward other bodies whose physical and behavioral traits give cues that they are harboring similar genes. In short, their inner selves may whisper, "Be nice toward those that resemble yourself." But at the same time, this angelic advice offered to one ear may be matched by a more subversive suggestion, whispered by a counterbalancing evolutionary devil perched on the other shoulder: "Be nasty toward those who are different."

This, then, is the ugly underbelly of kin selection: not selfishness, but racism, a special form of intolerance toward others, those who are biologically different, or, if nothing else, who look that way.

Although some people claim that the various human races are socially constructed and thus, biological fictions, the reality is otherwise. To be sure, there is no simple answer to the question, "How many races are there?" or "Are such-and-such a distinct race?" And there is absolutely no doubt that all human beings are members of the same species. It is also evident that the genetic differences between human races are biologically trivial, constituting less than 1/10 of one percent of total genetic makeup. But it is also clear that Caucasians, for example, are easily recognized as distinctly different from Chinese, and that either group is different from black Africans. Moreover, there can be no doubt that such differences—superficial as they are—reflect genetic differences: After all, black parents produce black offspring, pink parents produce pink offspring, and so forth.

Let me be clear: This is not to say that "race" is a particularly meaningful characteristic, nor is there any way that the human races can be in any way ranked as better or worse, superior or inferior. Moreover, nearly all of the differences among the races are more apparent than real; there is more genetic diversity, for example, among black Africans than between Caucasians and Asians. Nonetheless, racial traits exist, just as eye color exists, along with earlobe shape or blood type, and at least some of the differences among the races result from differences in their genes. This recognition, although it may make some well-meaning people uncomfortable, is demanded by old-fashioned intellectual honesty.

Those physical traits that characterize the various human races are the relics of genetically isolated groups of people (tribes) who remained isolated for many generations. Australian aborigines evolved kinky hair, residents of the Mongolian steppe evolved eyefolds, and so forth. Geography was presumably the cause of this genetic isolation. Even as the races came increasingly into contact, interbreeding has been limited by cultural traditions which have generally kept individuals from marrying far outside their social/biological group.

What does this have to do with kin selection, or with racism? Just this. As we have seen, human beings—like other living things—may well be predisposed to behave benevolently toward close relatives over distant relatives, and to favor distant relatives over strangers, at least in part because the closer the relative the higher the probability that genes will be shared. When it comes to recognizing one's kin, it seems highly likely that physical similarity has long been important: Everyone knows that relatives tend to resemble each other. And conversely, the less the resemblance, the less the likelihood of a close genetic relationship.

Skin color, eye shape, hair texture, physical size, nose shape, and other phenotypic differences among human beings reflect different ancestries. In general, the more differences, the more distant the genetic relationship. And the more differences, one can predict with some dismay, the less altruism.

The result may well be that human beings are naturally inclined—as a regrettable consequence of kin selection—to behave nonaltruistically toward others whose physical traits mark them as truly Other, that is, unlikely to be closely related. Once again, since this issue is so fraught with emotion and the potential for misunderstanding, let's be as clear as possible: Racism is in no way rendered acceptable just because it may be, to some extent, "natural." To the contrary, it is a practical and moral wrong that human beings are obligated to struggle against. But the fight against racism is not abetted by ignorance as to its possible origin.

Ironically, those racial categories that appear so prominent to so many people evidently reveal our tendency to establish social categories far more than they reflect biological reality. Nonetheless, human beings are acutely sensitive to the details of "exterior packaging" by which we identify each other as family, friends, or foe. It may be a tragic paradox that in unconscious pursuit of kin-selected benefits, we have come to exaggerate the significance of superficial differences that are just that: superficial.

In the musical *South Pacific*, a Caucasian lieutenant falls in love with a Polynesian woman. Reacting angrily to the racism of his society, he laments that society seems to demand that racism be taught, while children are still young and impressionable. Racism undoubtedly *can* be taught, and regrettably, it often is. So, fortunately, can racial tolerance and compassion. The point is that to some extent—exactly how far is unknown—people may indeed *have to be taught* tolerance, because left to their own devices, the whispers of kin-selected genes within most people seem to predispose them to a degree of bigotry that our species cannot afford.

We are children of the same mother—evolution—all of us nourished by the earth's good juices, yet our genes may well be programmed to see only narrower distinctions. To transcend ourselves, and our genes, is the uniquely human prerogative, as well as, increasingly, our responsibility.

"A thousand anachronisms dance down the strands of our DNA," wrote Barbara Kingsolver in *High Tide in Tucson*, a collection of her essays. "If we resent being bound by these ropes, the best hope is to seize them out like snakes, by the throat, look them in the eye, and own up to their venom."

17

Why Is Violence Such a "Guy Thing"?

WHEN IT COMES TO HUMAN VENOMOUSNESS, FEW THINGS ARE more toxic, or more biological, than violence. At the same time, although violence itself demands our attention, something about it—its pervasive *maleness*—is almost invisible.

Imagine that you were interviewing an intelligent fish, and you asked it to describe its environment. One thing it probably would *not* volunteer is that things are awfully wet down here! Like our hypothetical piscine interlocutor, people are generally insensitive to whatever permeates their lives. So, if you were to ask someone to describe human violence, only rarely would you hear that it is overwhelmingly perpetrated by males. And yet, the truth is that if we could eliminate—or even significantly reduce—male violence, we would pretty much get rid of violence altogether. The maleness of violence is so overwhelming that it is rarely even noticed; it is the ocean in which we swim.

What might be called the "killing establishment"—soldiers, executioners, hunters, even slaughterhouse workers—is overwhelmingly male. Underworld killers such as violent gangs are also peopled largely by men. Whenever seemingly unprovoked and deadly shootings occur in homes and workplaces, men are typically the mass murderers. Nor is this imbalance limited to the United States: whether in Kosovo, Rwanda, Cambodia, the Middle East, Guatemala, or Afghanistan, when people kill and maim other people, *men* are nearly always the culprits. And of course, the lethal operatives of Al Qaeda and equivalent are reliably male, as are those sent to combat them.

The same gender imbalance applies to the uncountable private episodes of violence that receive little national attention but are the stuff of many a personal tragedy. Admittedly, an occasional Lizzy Borden and her ilk surface, but for every Bonnie, there are about a hundred Clydes. Male brutalizers and killers are so common, they barely make the local news, whereas their female counterparts achieve a kind of fame. A man who kills—even his own children—gets comparatively little notice, whereas when Susan Smith drowned her two sons in 1994, she received international attention. For a man to generate a comparable response, his crime must be especially dreadful, such as serial murderer Ted Bundy or cannibal Jeffrey Dahmer, or a celebrity, like O. J. Simpson. Violence may or may not be as American as cherry pie, but it is as male as can be.

Violence is also, by and large, something that men direct toward other men. As with inner-city crime, in which both the perpetrators and victims are disproportionately minorities, men are disproportionately both the perpetrators and the victims of their own violence. This is not intended to romanticize or idealize women, or to deny that they too can sometimes be nasty, brutal, even deadly. Some women are more violent than some men, just as some women are taller, stronger, and have deeper voices and less hair than some men. But the overall pattern is consistent: when it comes to violence, the two sexes simply are not in the same league.

The same pattern is found, by and large, in animals, too.

Until a decade or two ago it appeared that other animals—including monkeys—did not kill members of their own species, whereas humans did. But as field studies in animal behavior have become more thorough, the myth of the peaceful primate—or nonmurderous animal generally—has largely been dispelled. Orangutans rape, for instance, and chimpanzees murder. Wolves also kill others of their own kind, as do lions, elk, and bison. In fact, nearly every animal species that has been carefully studied sooner or later reveals its penchant for lethal violence. And, to repeat, when such things take place among animals, the perpetrators—as among human beings—are almost always males.

Why is this? Evolutionary biology has an answer, and it emanates directly from the very definition of male and female.

* * *

Just look at the exterior genitals of a bird. In nearly every species, there aren't any! Males and females simply have a cloaca, the common external opening for excretory and reproductive products. And yet, biologists have no difficulty identifying male birds as distinct from females; ditto for male and female throughout the natural world. The difference between the sexes has nothing to do with penises or vaginas, beards or breasts. Rather, it is a matter of gametes: the tiny sex cells identifiable as either eggs (if large and produced in small numbers) or sperm (if small and produced in large numbers). This and only this is the "meaning" of maleness and femaleness: sperm-makers are called males, egg-makers, females.

The consequences of this distinction are weighty indeed. In brief, since sperm can be made in vast quantities, and with little mandated physiological follow-through, it is possible for males to have large numbers of offspring, the actual output limited by the number of females they succeed in fertilizing. By contrast, females are more likely to maximize their reproduction by producing successful offspring, rather than by outcompeting other females for the sexual attention of males.

To some degree, sexual competition is a replay of fertilization itself, in which numerous males, like hyperactive spermatozoa, compete among themselves for access to females. Just as it is now clear that the egg doesn't merely passively receive suitors, it is increasingly understood that females can be active participants in their own reproduction. Nonetheless, when it comes to sperm-makers, success is likely to crown those who outcompete their rivals, and so, in species after species, it is the males who are larger, nastier, more likely to be armed with lethal weaponry and a violent disposition to match. Natural selection has outfitted males with the perquisites for success in male-male competition, much of it violent.

In the animal world—human no less than nonhuman—competition is often intense. Males typically threaten, bluff, and if necessary fight each other in their efforts to obtain access to females. Among vertebrates in particular, males tend to be relatively large, conspicuous in color and behavior, endowed with intimidating weapons (tusks, fangs, claws, antlers, etc.) and a willingness to employ them, largely because such traits were rewarded, over evolutionary time, with enhanced reproductive success. A male with large tusks, for example, is more apt to win in battle against a lesser-endowed individual. Therefore, he will end up with access to more females, which in turn means that he will produce more offspring, and these offspring will likely have large tusks as well as their daddy's feisty disposition. Compared to females, as a result, males tend to be large, fierce, nasty, sneaky, and highly adapted to outmuscle, outshine, and occasionally, even outwit their rivals.

Male-male competition is especially fierce in polygynous, harem-keeping species such as elk, moose, elephant seals, or gorillas. Whereas in such cases each egg-maker is likely to be modestly successful (with one pregnancy per year), males play for higher stakes. They end up as harem-master or as an evolutionary failure, and not surprisingly, they grow up to become large, tough, and well-armed: unpleasant bullies as befits a winner-take-all lifestyle.

Consider elephant seals, behemoths that congregate annually to breed on islands off the coast of California. They are highly polygynous, with successful harem-keepers fathering upward of 40 offspring per year. And not surprisingly, the male elephant seal is truly elephantine, outweighing the female fourfold; he is also strongly disposed toward violence, nearly all of it directed toward other males. Why? Because among his ancestors, success has been rewarded . . . forty times per year. This, in turn, led to his great size as well as his inclination toward violence, neither of which is shared by the females, whose concerns are more intimately bound up with assuring the success of their offspring.

A bull elephant seal harem-master may have 40 mates, each of which will likely have a single pup. At the same time, since the sex ratio is one-to-one, for every harem-master, there will be 39 disappointed bachelors; as a result, some males will be immensely successful and others will be failures, while by contrast, the difference between success and failure is much less extreme among females. Think of it as different degrees of reproductive democracy, or egalitarianism. The payoffs to females are more equitable than that for males: one female, one offspring. Males, by contrast, operate within a system that is more inherently unfair and unequal. For them, there is a greater difference between the reproductive "haves" and the "have-nots." Hence, males are much more competitive than females.

In species that are monogamous or nearly so—such as most songbirds, geese, eagles, foxes, or gibbons—males and females produce approximately equal numbers of offspring. Not surprisingly, in such cases the two sexes are also nearly equal in physical size, armament, and aggressiveness. As we come to species that are more polygynous, however, we find a steady progression toward greater inequality in size and aggressiveness, with males getting bigger, and more nasty to each other. Among polygynous primates, for example, we find noticeable size differences between male and female, and also marked differences in behavior, especially when it comes to violence. A similar pattern holds for the deer family, the seals and their relatives, and indeed, pretty much any animal group that is diverse enough to permit comparisons of this sort. In addition, the greater the difference in reproductive payoff (variance in numbers of offspring), the greater the difference in aggressiveness among males. With reproductive success more variable, males are more competitive.

In most cases, noncompetitiveness among males results in nonreproduction. Consider the famous children's story of Ferdinand the Bull, a physically impressive creature who preferred smelling the flowers to fighting with other bulls. This "Ferdinand Option" simply does not exist for most males in most species, because Ferdinand would be less likely to promote himself—or, more to the point, his nonviolent, flower-sniffing genes—into the future. If Ferdinand has no heart for male-male competition, it may not matter if he has the testicles for it. Without willingness and ability to compete, his sperm—and thus his preference for flowers over fighting—are likely to be replaced by those of his less docile rivals.

This is not to deny recent findings that animals—even males!—often cooperate. My point is simply that because of the basic biology of maleness and femaleness, of sperm and eggs, males are more violence-prone. Incidentally, it had long been thought that the egg/sperm dichotomy also generated profound male/female differences in sexual proclivities. Even though recent DNA studies have revealed that females are more prone to sexual adventuring than had previously been thought, when it comes to violence, the male/female divide is as robust as ever.

As to basic reproductive biology, human beings are pretty ordinary mammals. *Homo sapiens* are also typically mammalian in their predisposition to polygyny (the mating system in which a successful male mates with numerous females); thus, our situation is consistent with that of elephant seals, although less extreme. Of 849 societies examined in anthropologist George P. Murdock's classic *Ethnographic Atlas*, 709 were polygynous, 136 were monogamous, and only 4 were polyandrous (one woman with many husbands).

Time and again, and regardless of the methodology used to obtain their sample, anthropologists have come up with similar results: Prior to the cultural homogenization that came with Judeo-Christian colonial—and marital—doctrine, polygyny was the preferred domesic system for more than 80 percent of human societies. (At the same time, even in non-Western, traditional cultures, most men did not actually succeed in becoming polygynists; monogamy, however, was nearly always imposed by necessity—usually poverty, personal inadequacy, and a shortage of potential mates—not choice.)

A Martian zoologist, reporting on the species *Homo sapiens*, would have no doubt: Human beings are mildly polygynous by nature. Like other polygynous mammals, we exhibit all the hallmarks: (1) males are typically larger and more aggressive than females; (2) females become sexually mature earlier than males; and (3) males have higher mortality rates, more rapid senescence, and shorter life spans.

Larger size and heightened aggressiveness were likely to lead to larger numbers of surviving children, especially in the long evolutionary childhood of the human species. As to age at sexual maturation, individuals of the more competitive sex nearly always mature later, thereby avoiding violent competition when their youth makes it adaptive for them to leave the breeding (and the serious fighting) to the older guys. Finally, the sex experiencing greater competition typically suffers higher mortality as a result. When these characteristics appear in other species, biologists readily interpret them as indicating male competition for access to females. Combined with the overwhelming cross-cultural data on *Homo sapiens*, we can safely conclude that in their history, human beings were polygynous. In their biology, they still are.

Hans Morgenthau, one of the great figures in 20th-century political science, used to argue that politics was based on male competition for power, a competition that this was, in turn, driven by three urges: to live, to propagate, and to dominate. Correct as far as he went, Professor Morgenthau might have been interested to learn that the first and third urges he identified are themselves proximate means to the middle one, the one that counts biologically: propagation. Reproduction, after all, lies at the root of *why* living things live, and *why* they seek to dominate. The ultimate power of propagation explains why males in particular are often so eager to dominate, occasionally carrying their eagerness to violent extremes. We should not be surprised to find that aggressiveness is widely—and all too correctly—seen as manly and its alternative, timidity, as womanly. (When told that a high-ranking member of his administration had become a dove on Vietnam, Lyndon Johnson snarled "Hell, he has to squat to take a piss.")

This is not to claim that females aren't aggressive in their own way. There are interesting cases of vigorous female-female competition in animals: Among groove-billed anis (large, ravenlike neotropical birds), several females deposit eggs in a communal nest, and the dominant female is especially likely to evict the eggs of subordinates; dominant female African hunting dogs may kill the offspring of lower-ranking females; female red howler monkeys push around other females. And in fact, many cases of monogamy among mammals may actually be enforced by subtle aggression on the part of females toward other females. I predict, in fact, that further research will reveal that female-female competition among animals is more widespread than currently realized. There is no doubt, however, that it is typically less direct, less boisterous, and much less violent than male-male competition. Female-female competition simply does not hold a candle to the brutal, bloody violence that so often characterizes competition among men.

On the domestic front, violent crime is overwhelmingly male. Studies of prosecution and imprisonment records in Europe, going back several centuries, as well as examinations of modern crime statistics from the United States and around the world, show that men consistently outstrip women when it comes to crimes by a ratio of at least three or four to one. When it comes to violent crimes, this difference is even greater, with the disparity increasing as the violence in question is more severe (simple assault versus assault and battery versus manslaughter versus homicide). The only areas, in fact, in which women commit more crimes than men are prostitution (which some would argue is not a criminal activity but an act between consenting adults) and shoplifting.

Another difference is that when women are consistently aggressive, it tends to be a defensive form, as when a woman kills a man who abuses her or her children, or fights to have a murderer condemned to death. It is interesting that among animals as well, a mother bear with cubs, for example, is notoriously fierce, as are other females who defend their young. Thus, although the aggression of women tends to be reactive, men are more likely to initiate violence, to commit truly "offensive" acts.

Although my concern here is with ultimate, evolutionary answers, this does not preclude noting that the male sex hormone, testosterone, is associated with violent crime among men, although the correlation is not simple. Interestingly, high testosterone levels are also correlated with violent crime among women. In short, although women commit fewer violent crimes than men, those who do so appear to have higher circulating levels of testosterone than those who are less violent. To put it simply, women carrying a "macho" dose of hormones seem more likely to be violent; that is, to be like men.

Although the precise mechanism remains obscure, at present we also know that the sexes are not equally vulnerable to mental illness. Women are more likely to suffer from depression. Men are more vulnerable to certain mental illnesses that are correlated with violence. For example, adolescent boys greatly exceed girls among those suffering from the descriptively labeled "oppositional defiant disorder." When it comes to "general conduct disorders," the prevalence among males under age 18 ranges from 6 percent to 16 percent; for females, 2 percent to 9 percent. Two to five times more males than females are heavy drinkers, and for a wide range of "impulse control disorders," including intermittent explosive disorder, pathological gambling, and pyromania, males greatly exceed females. It is almost—but not quite—comical to note that women exceed men in kleptomania and trichotillomania. (That is, when men express less impulse control than women, the results are likely to be violent; when women express less impulse control than men, it is to shoplift, or pull out their hair.)

When it comes to the most serious violent crime, homicide, men are far and away the most frequent perpetrators. Interestingly, they are also most likely to be the victims, precisely as evolutionary theory predicts. Thus, murder is largely a crime of men against other men, a finding that, in itself, points an accusing finger at male-male competition. In their book *Homicide*, Canadian psychology professors Martin Daly and Margo Wilson reviewed murder records, specifically looking at cases involving members of the same sex, over a wide historical range and from around the world. They concluded that "the difference between the sexes is immense, and it is universal. There is no known human society in which the level of lethal violence among women even begins to approach that among men."

Daly and Wilson found that a man is about 20 times more likely to be killed by another man than a woman is by another woman. This holds true for societies as different from each other as modern-day urban America (Philadelphia, Detroit, and Chicago), rural Brazil, traditional village India, Zaire, and Uganda. This is not to say that actual murder rates are the same in these different places. In modern Iceland, for example, 0.5 homicides occur per million people per year, whereas in most of Europe, the figure is closer to 10 murders per million per year, and in the United States, over 100. The crucial point is that despite these wide differences, the basic male-female pattern remains stable: male-male homicide exceeds its female-female counterpart by a whopping margin. The fact that the ratio of male-male to female-female violence remains remarkably *un*varying from place to place argues for its biological underpinnings and parallels the male-male competition seen in other species.

The same trend can be found across history. Thus, even though a 13th-century Englishman was 20 times more likely to be murdered than an Englishman is today, he was 20 times more likely to have been murdered by another *man* than an Englishwoman was by another woman. Not only that, but around the world and across time, the age of the vast majority of these male murderers remains constant, in their mid-20s. (Think of newly mature male elephant seals, challenging for a place at the reproductive table.)

While in recent years women have been increasingly involved in criminal behavior, Daly and Wilson cite FBI statistics attributing this increase to growing numbers of women arrested for "larceny-theft," whereas the proportion of women arrested for violent crimes—and for homicide in particular—has actually declined slightly.

In 1958, sociologist Marvin Wolfgang published what has remained the classic study of homicide in America, based on nearly 600 murders in the city of Philadelphia. Trying to explain why more than 95 percent of the killers were men, Wolfgang—a proponent of learning theory and cultural explanations—wrote: "In our culture [the average female is] . . . less given to or expected to engage in physical violence than the male." We are supposed to infer that things are different in other cultures, but this simply is not so.

There is a powerful bias in the United States, promoted by most contemporary psychologists, anthropologists, and sociologists, that male-female differences have been created solely by differences in upbringing and social expectations. As a result—whether by error or preexisting bias—social scientists have contributed to a vast myth: the myth of the equipotential human being, the idea that everyone is equally inclined to behave in any which way. Equipotentiality is an appealing sentiment, attractively egalitarian. There is only one problem: It isn't true. Quite simply, it flies in the face of everything known about the biological underpinnings of behavior, and of life itself.

Moreover, if male-female differences derived essentially from arbitrary cultural traditions—the well-known phenomenon in which societies typically imbue young men with the expectation of greater violence—there should be at least some in which the situation is reversed, where young women are socialized to be the more violent sex. But there aren't any.

Male-male competition doesn't affect only those who are successful. Indeed, males can be as ferocious trying to avoid total defeat (seeking *not* to be the 40th elephant seal) as they are when trying to rise to the top of the heap. Not infrequently, battles at the lower end of the competitive ladder are even more vicious than those among the elite. This is because men at the bottom of the sociosexual hierarchy have little or nothing to lose, and so are especially likely to fight no-holds-barred, with a kind of last-ditch bravado that uses the riskiest and most deadly tactics.

Violence is often seen as primitive or immature. And yet, the reality is that even in this era of gun-toting 12 year olds, murderous violence is distressingly mature: overwhelmingly, it is *adult* behavior. It is also easily triggered. When Marvin Wolfgang conducted extensive interviews with convicted killers in Philadelphia, he was able to identify 12 different categories of motive. Far and away the largest, accounting for fully 37 percent of all murders, was what he designated "altercation of relatively trivial origin; insult, curse, jostling, etc." In such cases, people got into an argument at a bar over a sporting event, who paid for a drink, an offhand remark or a hastily uttered insult, etc.

To die over something so inconsequential as a casual comment or a dispute about some distant event seems the height of irony and caprice. But in a sense, disputes of this sort are not trivial, for they reflect the evolutionary past, when personal altercations were the stuff upon which prestige and social success—leading ultimately to biological success—were based. It is not surprising, therefore, that young men today will fight and die over who said what to whom, whose prestige has been challenged, and so forth.

Within a group subject to discrimination, the pressures and pains—as well as the tendency to "act out"—will be especially strong. Another way to look at it: the fewer the opportunities for social success, the greater the risks worth taking. From an evolutionary perspective, therefore, it is not surprising that it is young *men*, especially from disadvantaged social and ethnic groups, who are overrepresented in drug addiction, violent crime, penitentiaries, and death row. And that angry and alienated *men* comprise the overwhelming majority of violent terrorists.

Others have tried to explain the high rate of male violence without regard to biology. For example, advocates of social learning theory point out that men—whether African American, Caucasian, Asian, or whatever—are *expected* to be aggressive; women are *supposed* to be more passive, etc. So people grow up this way, it is claimed, meeting the expectations that society imposes on them. But why should society have such expectations? And why are those expectations virtually the same in every society around the world? And why do both men and women find it so easy to comply?

British psychologist Anne Campbell, an advocate of social learning and cultural influence, thinks that men are more aggressive than women because men and women interpret aggression differently: women see it as a loss of self-control, and are ashamed of their anger, associating it with being pushy, nasty, socially isolated. Men, by contrast, see their aggressiveness in a positive light, as a way of *gaining* control. To men, anger and even rage can mean courage, success, and triumph. Campbell's analysis is probably correct as far as it goes. But why do males associate aggression with success? And why do they view controlling others as more important than controlling themselves? Also, why do women feel so threatened by isolation and anything that smacks of diminished intimacy, while men feel threatened by anything that smacks of diminished prestige and authority? If the "answer" is that women are taught to react as they do, then I must repeat: Why are virtually identical patterns found in every culture on earth? And why are similar patterns even found in the most different "cultures" of all, those of other species?

All of the above is not meant to imply that biology is the sole explanation for the gender gap in human violence. We cannot do a thing about our evolutionary bequeathal; hence, we had better do all we can to ameliorate those conditions that predispose people to violence. And let's face it: Biology does in fact explain a whole lot, such that if we are going to intervene effectively we would be well advised to understand the nature of our own predispositions. Just like our make-believe finny friend with which this essay began, it is time for all of us to look around and acknowledge that when it comes to the "social construction" of sex differences in violence, the traditional view is all wet.

18

One and a Half Cheers . . .

WOMEN, OF COURSE, CAN ALSO BE AGGRESSIVE, AND NASTY, AND cruel (also—like men—they can be pacific, and kindly, and gentle). Although evolutionary biology helps us understand why men are typically more violent than women, neither sex has a monopoly on war versus peace; similarly, and contrary to widespread assumption, human beings do not have a monopoly on violence.

I well remember an exhibit at the Bronx Zoo when I was a child (it has since been copied by zoos throughout the world). It offered a view of the "world's most dangerous creature," and was, of course, a mirror. No reasonable person—least of all anyone with environmental sensibilities—can doubt the veracity of this assertion, intended to shock the zoo-goer into a healthy degree of eco-friendly self-reflection. Nor can anyone doubt that human beings are not only dangerous to their planet and many of its life-forms, but, most of all, to themselves.

Homo sapiens has much to answer for, including a gory history of murder and mayhem perpetrated upon one another. Anthropologist Raymond Dart spoke for many when he lamented that "the atrocities that have been committed . . . from the altars of antiquity to the abattoirs of every modern city proclaim the persistently bloodstained progress of man." An unruly, ingrained savagery, verging on blood-lust, has been a favorite theme of fiction, including, for example, Joseph Conrad's *Heart of Darkness* and William Golding's *Lord of the Flies*, while Robert Louis Stevenson's *The Strange Case of Dr. Jekyll and Mr. Hyde* developed an explicit notion of duality: that a predisposition to violence lurks within the most outwardly civilized and kindly person.

There even seems to be a curious, Jekyll and Hyde–like ambivalence in humanity's view of itself: on the one hand, we have Protagoras's insistence that "man is the measure of all things," linked theologically to the biblical claim that "God made man in his own image." The upshot: Human beings are not only supremely important, but may be even supremely good. At the same time, however, there is another, darker perspective, promoted not only by environmental educators, but also certain Christian theologians as well as nonsectarian folks who so love humanity that they hate human beings . . . largely because of what these same human beings have done to other human beings.

In extreme cases, the result has been outright loathing, often stimulated by conviction that humanity is soiled by original sin and is, moreover, irredeemable, at least this side of heaven. According to the zealous John Calvin,

> The mind of man has been so completely estranged from God's righteousness that it conceives, desires, and undertakes, only that which is impious, perverted, foul, impure and infamous. The human heart is so steeped in the poison of sin, that it can breathe out nothing but a loathsome stench.

Misanthropy can also be purely secular, as in this observation from Aldous Huxley:

> The leech's kiss, the squid's embrace,
> The prurient ape's defiling touch:
> And do you like the human race?
> No, not much.

In a similar vein, human beings stand accused of being not only murderous but uniquely so, an indictment that has been largely transformed into a guilty verdict, at least in much of the public mind. Writing in 1904, William James described "man" as "simply the most formidable of all the beasts of prey, and, indeed, the only one that preys systematically on its own species." A half-century later, this view was endorsed by no less an authority than pioneering ethologist and Nobel Prize–winner Konrad Lorenz, who popularized the idea that lethally armed animals (wolves, hawks, poisonous snakes) are also outfitted with behavioral inhibitions that prevent their use against conspecifics. Human beings emerge as the sole exception, since our lethality is "extrabiological," rendering us anomalous in our uninhibited murderousness. Paradoxically, such claims have been widely—and even warmly—embraced. "Four legs good, two legs bad," we eagerly learned from George Orwell, not least because *Homo sapiens* is supposed to be uniquely branded, among all living things, with the mark of Cain.

There appears to be a certain pleasure, akin to intellectual self-flagellation, that many people—college students, it appears, most especially—derive in disdaining their own species. Maybe anathematizing *Homo sapiens* is a particularly satisfying way of rebelling, since it entails enthusiastic disdain of not merely one's culture, politics, and socioeconomic situation, but one's species, too. At the same time, such a posture is peculiarly safe, because species-rejecting rebellion does not require casting aside citizenship, friends and family, or access to one's trust account: Having denounced one's species, nobody is expected to join another.

In any event, Cain is a canard. We have no monopoly on murder. Human beings may be less divine than some yearn to think, but—at least when it comes to killing, even war—we aren't nearly as exceptional, as despicably anomalous and aberrant in our penchant for intraspecies death-dealing, as the self-loathers would have it.

The sad truth is that many animals kill others of their kind, and as a matter of course, not pathology. When anthropologist Sarah Hrdy first reported the sordid details of infanticide among langur monkeys of India, primatologists resisted the news: it couldn't be true, they claimed. Or if it was, then it must be because the monkeys were overcrowded, or malnourished, or otherwise deprived. They couldn't possibly stoop to killing members of their own species (and infants, to make matters even worse); only human beings were so depraved. But in fact, this is precisely what they do. More specifically, it is what male langur monkeys commonly do when one of them takes over control of a harem of females: The newly ascendant harem-keeper proceeds, methodically, to kill any nursing infants, which, in turn, induces the previously lactating (and nonovulating) females to begin cycling once again. All the better to bear the infanticidal male's offspring, my dear.

We now know that similar patterns of infanticide are commonplace among many other species, including rats and lions, as well as other nonhuman primates. In fact, when field biologists encounter a "male takeover" these days, they automatically look for subsequent infanticide and are surprised if it *doesn't* occur.

The slaughter of innocents is bad enough (by human moral standards), although not unknown, of course, in our own species. But from a strictly mechanistic, biological perspective, it makes perfect sense. It might also seem more "justifiable" than, say, adults killing other adults, if only because the risk to an infanticidal male is relatively slight (infants can't do much to defend themselves), and the evolutionary payoff is comparatively great: getting your genes projected into the future via each bereaved mother, who would otherwise continue to nourish someone else's offspring instead of bearing your own. (More accurately: genes getting themselves projected into the future via. . . .) In any event, the evidence is overwhelming that among many species adults kill other adults, too.

Lorenz was right, up to a point. Animals with especially lethal natural armaments tend, in most cases, to refrain from using them against conspecifics. But not always. In fact, the generalization that animals—predators and prey excepted—occupy a peaceful kingdom was itself greatly overblown. Maybe some day the lion will lie down with the lamb, but even today lions sometimes kill other lions and rams knock down (thereby knocking off) other rams. The more hours of direct observation biologists accumulate among free-living animals, the more cases of lethality they uncover. Indeed, a Martian observer spending a few weeks among human beings might be tempted to inform his colleagues, with wonderment and some admiration, that *Homo sapiens* never kills conspecifics. She would be as incorrect as those early reports that wolves invariably inhibit lethal aggression by exposing their necks, or that chimpanzees make love instead of war.

In fact, wolves do kill other wolves, showing little mercy for outliers and other strangers. And chimpanzees make war.

To be sure, if one defines war as requiring the use of technology, then our chimp cousins aren't warmongers after all, but if by war we mean organized and persistent episodes of intergroup violence, often resulting in death, then chimps are champs at it. Jane Goodall has reported extensively on a four year–running war between rival troops of chimpanzees in Gombe National Park in Tanzania. And similar accounts have emerged from other populations in Budongo and Kibale Forests in Uganda, Mahale Mountains National Park in Tanzania, and Taï National Park in Côte d'Ivoire. Chimpanzee wars are not an aberration.

As to why they occur, anthropologist Richard Wrangham explains that "by wounding or killing members of the neighboring community, males from one community increase their relative dominance over their neighbors. . . . This tends to lead to increased fitness of the killers through improved access to resources such as food, females, or safety." These episodes typically involve border patrols leading to organized attacks in which a coalition (composed almost exclusively of males) will attack, and often kill, members of the neighboring troop (once again, almost exclusively males).

At this point, some readers—struggling to retain the perverse pride that comes from seeing human beings as, if not uniquely murderous, then at least unusually so—may want to backpedal and point out that chimps are, after all, very close to *Homo sapiens*. But in fact, lethal fighting—albeit less organized than chimpanzee warfare—has been identified in hyenas, cheetahs, lions, and many other species. In one study, nearly one-half of all deaths among free-living wolves not caused by humans were the result of wolves killing other wolves. *Homo homini lupus*? Indeed, but with this caveat: wolves, too, are wolves to other wolves.

And so are ants. According to Edward O. Wilson, America's supreme ant-ologist, "alongside ants, which conduct assassinations, skirmishes, and pitched battles as routine business, men are all but tranquilized pacifists." In their great tome of ant lore, Wilson and Bert Hölldobler concluded that ants are "arguably the most aggressive and warlike of all animals. They far exceed human beings in organized nastiness; our species is by comparison gentle and sweet-tempered." The ant lifestyle is characterized, note Wilson and Höll-dobler, by "restless aggression, territorial conquest, and genocidal annihilation of neighboring colonies whenever possible. If ants had nuclear weapons, they would probably end the world in a week."

Primatologists Alexander Harcourt and Frans de Waal (the latter having written extensively about "natural conflict resolution," and, if anything, predisposed to acknowledge the pacific side of animals) conclude that, regrettably but undeniably, "lethal intergroup conflict is not uniquely, or even primarily, a characteristic of humans." The bottom line: Our species is special in many ways, and we may even be especially accomplished when it comes to killing our fellow humans, but insofar as same-species lethality goes, we are not alone.

Jonathan Swift was no sentimental lover of the human species, verging—and sometimes settling—on outright misanthropy. Thus, during one of Gulliver's voyages, the giant king of Brobdingnag describes human beings as "the most pernicious race of little odious vermin that nature ever suffered to crawl upon the surface of the earth." Swift himself wrote, "I hate and detest that animal called Man, yet I heartily love John, Peter, Thomas and so forth." It is Gulliver's final voyage, however, to the land of the admirable, rational, equably equine Houynhnms that constitutes what is probably the most sardonically critical account of humanity, in all its Yahoo nature, ever written. Sir Walter Scott wrote that this work "holds mankind forth in a light too degrading for contemplation."

Especially degrading—for Swift, Scott, and, as the story unfolds, the Master of the Houynhnms—is the human capacity for lethal violence, especially during war:

> [B]eing no stranger to the art of war, I [Gulliver] gave him a description of cannons, culverins, muskets, carbines, pistols, bullets, powder, swords, bayonets, battles, sieges, retreats, attacks, undermines, countermines, bombardments, seafights; ships sunk with a thousand men; twenty thousand killed on each side; dying groans, limbs flung in the air: smoke, noise, confusion, trampling to death under horses feet: flight, pursuit, victory, fields strewed with carcasses left for food to dogs, and wolves, and birds of prey; plundering, stripping, ravishing, burning and destroying. And, to set forth the valour of my own dear countrymen, I assured him that I had seen them blow up a hundred enemies at once in a siege, and as many in a ship; and beheld the

> dead bodies drop down in pieces from the clouds, to the great diversion of all the spectators.

Omitted, for obvious reasons: machine guns, submarines, mustard gas, mechanized artillery, land mines, fighter planes, bombers, cluster bombs, nuclear warheads, and other weapons of mass destruction (and this is a woefully incomplete list), not to mention the use of commercial airliners as weapons of mass destruction, or the use of lies about weapons of mass destruction to justify an invasion that results in tens of thousands of unnecessary deaths.

Let's face it, human beings are a murderous lot, destructive of each other no less than of their environment. But let's also admit that such misdeeds, grievous as they are, derive less from a one-of-a-kind blood lust than from the combination of all-too-natural aggressiveness with ever-advancing technology—which is itself natural, too.

Tennyson was correct after all. Nature really is red in tooth and claw—not always, to be sure, but more often than a romanticized view of the animal world would have us believe. And not only when it comes to predators dispatching their prey. Also, not merely tooth and claw, but antler and horn and stinger and tusk and butcher knife and Kalashnikov. We aren't so much separated from nature as connected to it, for worse as for better, empowered by our culture to act—often excessively, because of the potent technological levers at our disposal—upon impulses that are widely shared. And so, one and a half cheers for *Homo sapiens*, the world's most dangerous creature, whose dangerousness resides not in the originality of its sin, but in the reach of its hands.

19

Honest Liars?

According to Mark Twain, human beings are the only animals that blush . . . or need to. He was presumably thinking of our numerous episodes of inhumanity to man, when, as the adage goes, *Homo homini lupus* ("man is a wolf to men"—something of a calumny upon wolves). As just argued in the last chapter, however, people are indeed violent and murderous, but not uniquely so. Maybe Mr. Twain was meditating, instead, on some of our other sins, such as dishonesty.

One cannot lie unless there is some possibility of telling the truth; in a world without truth-telling, lying is meaningless. And among the intriguing consequences of evolutionary thought has been its reformulation of the very idea of truth, lies, and—more fundamentally—communication. Educators are supposed to be experts at communication (at least, we are paid to do it). But what *is* communication? In this case, it won't do to quote Justice Potter Stewart on pornography—we may not be able to define it, but we know it when we see it—because, in fact, people often do not know communication when they see it. Or rather, they are often deceived about it. And that is precisely the point.

In the most traditional definition, communication is simply the transfer of information from a sender to a receiver. Also assumed: The sender benefits by sending and the receiver by receiving. The result, accordingly, would seem to be a situation of happy collaboration, in which any difficulties are mere testimony to how hard it can be to bridge the unavoidable gap between individuals. Anyone who readily believes this, however, has probably never tried teaching a class, or looking hard at how animals communicate with one another.

Let's not worry overmuch about what professors (or writers about evolution!) are trying to communicate, not only the precise subject matter but even whether the goal is to transmit facts or, more popularly, "how to think." Either way, information is to be transmitted. And yet, Summerhillian wishful thinking notwithstanding, the reality is that students don't always want to learn. Often, for example, they want to absorb only what is likely to be on the next test. "Are we responsible for this?" must be one of the most frustrating questions an instructor can ever hear. And in response, teachers often seek to confound student "receptor selectivity" by not revealing what, amongst the large amount of information being transferred, is going to be tested. Students, in turn, respond by trying to anticipate what is going to be tested. The result can be something less than shared interests, and more like a battle of wits. The old, cynical definition of a lecture—the exchange of information from the notebook of a professor to those of the students, without passing through the brain of either—may therefore have to give way to an equally cynical alternative, which emphasizes resistance on the part of the receiver and, often, an ulterior motive by the sender.

Here it might help to look at animal communication. In the heyday of so-called classical ethology, as pioneered by Niko Tinbergen and Konrad Lorenz, a cheerfully naïve view of communication held sway. Sender and receiver were presumed to be "on the same page," as animals indicated, for instance, their internal state (aggressive, defensive, sexually aroused, etc.) by various postures and behavior, and receivers were concerned only with decoding the messages as accurately as possible. In this dance of mutual benefit, the blue-ribbon winner has been a dance itself: the celebrated "dance of the bees," whereby a foraging worker, having discovered a good food source, communicates its location—with remarkable accuracy—to others within a darkened hive. There is no reason to suspect a dancing worker bee of dishonesty. But as evolutionary theorists eventually came to realize, other cases are likely to be much darker; there is trouble in this presumed communicative paradise.

Long ago, during one of his concerts, musical satirist Tom Lehrer noted that "if people are having trouble communicating, the least they could do is shut up about it." I have long admired Mr. Lehrer, but—in this regard at least—I respectfully disagree.

In *Henry IV, Part I*, Owen Glendower boasts, "I can call spirits from the vasty deep," to which Hotspur responds, "Why, so can I or so can any man; But will they come when you do call for them?" Under what conditions would those spirits emerge from the vasty deep? Hint: not simply because Mr. Glendower has called. Rather, an evolutionary perspective suggests that they should only come when it is in *their* interest—not Glendower's—for them to do so. After all, Glendower might want to harpoon them, or saddle them, or tickle them unmercifully for his own selfish gratification. On the other hand, maybe the spirits will profit by being called: perhaps the mystical Welshman has good news, something of value to share, or useful information to impart. Useful to the spirits, that is.

All receivers, whether spirit or organic, should accordingly be selected to discriminate self-serving from beneficial "calls" emanating from the likes of Glendower. (Which is to say, from anyone.) And insofar as he gains by inducing the spirits to respond to his communication, Glendower, in turn, should be selected to send messages that would appeal to the apparent self-interest of the recipients, whereas in reality, they more likely contribute to his own. In short, given the fundamental self-interest of the evolutionary process, communication may well be dishonest.

We are accustomed to deceit in animal communication when it operates between different species. After all, that's what camouflage, for example, is all about: A ptarmigan, white against the winter snow, says "I'm not here." A stick insect says "I'm a stick, not an insect." Other potential prey items claim to be a leaf, a bit of bird poop, the eyes of an owl that preys upon whatever might otherwise try to prey upon it. But what about communication within the same species?

In some cases, such as those marvelous dancing bees, genetic interests are closely shared. As a result, no deception is expected. But given that individuals and their genes are selected to maximize their own benefit, not that of others, we can expect that their interest in honesty and accuracy is rather limited. In fact, communication may be indistinguishable from manipulation. "When an animal seeks to manipulate an inanimate object," write Richard Dawkins and John Krebs,

> it has only one recourse—physical power. A dung beetle can move a ball of dung only by forcibly pushing it. But when the object it seeks to manipulate is itself another live animal there is an alternative way. It can exploit the senses and muscles of the animal it is trying to control, sense organs and behavioural machinery which are themselves designed to preserve the genes of that other animal. A male cricket does not physically roll a female along the ground and into his burrow. He sits and sings, and the female comes to him under her own power.

Consider, more generally, the case of a male wishing to convince a female that he is a suitable mate. Since the male is typically capable of copulating many times, it is often in his interest to persuade females to choose him over his competitors. This puts females in a position of having to discriminate among various suitors. They, in turn, are therefore likely to exaggerate and if necessary, misrepresent their quality as a mate, essentially sending this message: "Choose me, I'm the healthiest (or the strongest, or the best provider, and so forth)." Insofar as these representations depart from the truth, they are the animal equivalent of lying. Females, in turn, are selected to see through the false advertisement, leading to a mutual arms race: Greater deceit by the senders leads to enhanced ability to discriminate on the part of the receivers, which in turn leads to yet more devious deception, countered by yet more sophisticated discrimination, and so forth.

Nor is this dance of deception necessarily limited to the much-touted battle of the sexes. When different troops of vervet monkeys meet, someone (typically from among the low-ranking males) may well give an alarm call, of the sort normally reserved for when a leopard is on the scene. The result? Everyone heads for cover, and a violent intergroup clash—in which low-ranking males typically come out worst—is avoided. It is possible that such calls during intergroup encounters are simple mistakes. But this is unlikely, since vervets are remarkably sophisticated when it comes to alarm calling (for example, they employ three distinct calls to warn of aerial predators, ground predators such as leopards, and snakes). Moreover, it stretches credulity that only low-ranking males—those who profit from the act—should consistently be the ones making the same "mistakes."

Here is another example of manipulative alarm-calling. The paternity of male barn swallows is threatened by the prospect that females will engage in "extra-pair copulations" with other males. Alarm calls, not surprisingly, break up any such trysts, the risk of predation evidently being more salient than the allure of avian adultery. This led to a field study in which female barn swallows were chased from their nests while the male was out foraging. Upon his return, finding "his" female absent, male barn swallows gave alarm calls, which would likely disrupt any ongoing extra-pair copulation. Significantly, such dishonest signaling did not take place when females weren't fertile; moreover, it was only performed by colonial breeding swallows and not by solitary ones, whose males are not at risk of being cuckolded.

There are many other cases in which the honesty of communicators can be questioned. Consider two lizards confronting each other over a mutually desired resource such as a nest site, a mate, or a morsel of food. They communicate via threat displays, each seeking to induce the other to back down, thereby avoiding a potentially damaging battle. Success means gaining the resource without a fight, but whoever retreats gets nothing. Under such circumstances, it should pay each contestant not only to communicate its size, strength, and determination, but, if anything, to exaggerate these traits, if by doing so it is more likely to come out ahead. Or take a gazelle, being eyed by a cheetah: Since a predator is less likely to waste time and energy chasing prey that it cannot catch, it might well be in the interest of potential prey to communicate to a would-be predator that the latter has been seen and has therefore lost the benefit of surprise, as well as that this gazelle, at least, is so quick, agile, and healthy that pursuit would be unavailing.

But what is to stop a gazelle from sending a dishonest message, seeking to indicate that it is quicker, more agile, and healthier than it really is? (In which case, cheetahs would no longer take gazelle messages seriously, assuming that they ever did.)

A compelling answer has been proposed by Israeli zoologist Amotz Zahavi, who suggests that, indeed, communication is seriously bedeviled by the temptation to cheat and send dishonest signals. As a result, argues Zahavi, receivers are likely to pay special attention to messages that are inherently protected against cheating and are necessarily honest because they are expensive and difficult if not impossible to fake. Hence, gazelles that might otherwise be stalked by cheetahs engage in "stotting," a peculiar, high, stiff-legged jump that requires quickness, agility, and overall health. Sickly gazelles can't stot. Females of many species insist that courting males actually present them with genuine prey items, whose nutritive value cannot be faked, just as a male's quality as hunter, scavenger, or provisioner is also guaranteed to be real if he actually has a nuptial meal to offer.

More controversially, Zahavi's "handicap principle" suggests that the choosier sex (generally females) will be selected to prefer potential mates whose courtship proceeds despite their possession of a handicap, which serves as a guarantee of quality and thus of honest communication. Under this view, for example, females prefer to mate with males sporting gaudy secondary sexual characteristics (bright feathers, colorful wattles, elaborate and expensive courtship shenanigans) precisely because these traits, being significant handicaps, are very difficult to fake. A parasite-ridden, genetically challenged, or metabolically stressed suitor would be less able to manufacture the elaborate tail of a successful peacock, "boom" like an impressive sage grouse, carry a platinum charge card, or swim the Hellespont.

What, if anything, does all this mean for those of us struggling in the groves of academe (or in daily living) to—let's be honest, now—communicate? Would our lectures (or our more intimate communication) be more persuasive if we handicap ourselves, perhaps by proceeding without benefit of microphone, sans PowerPoint, or maybe while standing on one leg? Or if we strove, somehow, to show that all we desire is to impart information and, occasionally, our version of wisdom, selflessly, and "just to communicate" rather than to manipulate?

I once heard undergraduate lecturing defined as the casting of artificial pearls before genuine swine, whereupon I self-righteously objected that students (for the most part) aren't swinish and that, in any event, most lectures (at least, mine!) may not consist entirely of pearls, but are at at least genuine. But I was younger then, and more naïve. Now, I'm not so sure.

20

What Puts the Dys in Dystopia?

WITHOUT THE USUAL FANFARE, WE'VE ENTERED TERRITORY NOT often traversed by evolutionary biologists, and controversial terrain at that: the fraught phenomenon of language and communication, symbolic and cultural processes, often assumed to be uniquely human and occupied solely by "the humanities." Let's press on, taking note of some parallels between sociobiology and the creative arts, in particular some of the stories people tell themselves. Among those stories are utopias, imagined worlds in which things are better than in our own. Even more compelling, however, are dystopias, imagined worlds in which things are somehow worse. The evolutionary biologist can't help noticing that literary dystopias have this in common: They all involve societies in which the deepest demands of human nature are either subverted, perverted, or simply made unattainable. Not that it is necessarily bad to say "no!" to human nature. When it comes to certain inclinations, such as violence or extreme selfishness, there is much to be said for defying the promptings of biology. But when society presses too hard in ways that go counter to natural needs, the result can be painfully unnatural; which is to say, dystopian.

What are some exemplary dystopias? Foremost for many are Aldous Huxley's *Brave New World* and George Orwell's *1984*. The towering influence of this literary duo is due not only to their imaginative and artistic quality, but also to the powerful theme that all dystopian literature shares: the horror of a society that runs roughshod over our instincts, forcing people to be, literally, inhuman.

In Huxley's world, sex has been separated from reproduction: the former takes place quickly, easily, and without commitment or emotional involvement; the latter, in gigantic, highly technological Hatcheries wherein embryos are created and fertilized and babies "born." The horror of this society is so great that an outsider, "John the Savage," eventually kills his lover and hangs himself, in a frenzy over its lack of poetry, insensitivity to love, and indifference to death.

No outlet here for anything approaching a normal biological urge; in fact, the words "father" and "mother" are cause for scandal. The human need for affection is denied, and with it, much of human nature itself. The Director of Hatcheries describes any "emotional" and "longdrawn" interactions with the opposite sex as "indecorous," his disinterest in romance contrasting with the novel's title, which was inspired by these rapturous words of Miranda in Shakespeare's *The Tempest*, after she falls head-over-heels, *humanly* in love: "Oh brave new world, that has such people in't!" It is precisely this exultant, hormonally charged intoxication that is anathema in Huxley's *Brave New World*, where there are no parents to love children, or sons and daughters to return the sentiment. Indeed, there is no genuine love at all. In what many might perceive as a positive departure from human nature, sexual jealousy is also abolished, since "everyone belongs to everyone else." Yet love, sex, and jealousy are primal aspects of the human psyche; to deny them is to deny our biological selves.

Think, next, of George Orwell's *1984*. What comes to mind is mostly "Big Brother," "newspeak," and "thought control," as well as linguistic incongruities and the terrifying consequences of resistance. As a paradigm for statewide dehumanization, the Party's combination of brainwashing and ferocity is consummately successful. But perhaps most inhuman, and therefore most disturbing, about the state of Oceania is its routine undermining of social interactions. No wonder the hero, Winston Smith, is most sympathetic during his futile attempts to establish personal connections with a fellow human being.

Through the Party's obsession with "chastity and political orthodoxy," *1984* is almost a textbook account of how to organize *Homo sapiens* in ways that contradict their most basic biological needs. Not just sexual desire, but even genetic continuity is placed at risk: The prospect of staying alive through time via future generations is the motivation underlying sex, love, and indeed everything in the organic world; accordingly, Orwell's dystopia recognizes the biologically induced terror of genetic erasure. The Party's preferred response to opponents is simple elimination: "Your one-time existence was denied and then forgotten. You were abolished, annihilated: *vaporized* was the usual word." This is precisely what genes—experts as they are in self-perpetuation—do not want.

Social destruction in this antibiological dystopia includes even the elimination of basic sociality. Intrinsically a group-living ape, our species shudders at the prospect of being alone. The horror of the friendlessness experienced by Winston Smith—"you did not have friends nowadays, you had comrades"—is more deep-seated than simple longing: it is an expression of the elemental importance of social life itself. And, true to form, what should be the strongest of all social units—the family—is attacked the most severely. Big Brother's spies destroy the integrity of family, such that "it was almost normal for people over thirty to be frightened of their own children." To have one's own genes turn against one's self: Is there any greater potential perversion of the biological world?

In justifying this nightmare society, Winston's torturer, O'Brien, explains: "You are imagining that there is something called human nature which will be outraged by what we do and will turn against us. But we create human nature. Men are infinitely malleable." Fortunately, O'Brien, like the Director in *Brave New World*, is wrong. People are immensely malleable, more so, in all likelihood, than any other species. But *infinitely*? Absolutely not. And it is precisely such asserted distortions of biological reality that make *1984*, as with *Brave New World* before it, so deeply troublesome.

Denial of love, of genuine sex (which is to say, difficult, but also gratifying), of reproductive opportunity, of individuality: fundamentally, all are denials of our organic human-ness. One of the most powerful such representations comes from the early Soviet-era dissident writer Yevgeny Zamyatin, in his brilliant, chilling, pre-Huxley/Orwell dystopia, *We*. Life in Zamyatin's One State, orchestrated by the Great Benefactor, is carried on by numbers, not individuals. There are no primitive passions, no instincts; everything is designed with mathematical precision. Nature—which is both feared and hated—has been banished behind the Green Wall, which, as the narrator, D-503, explains, enables man to be no longer "a savage." Although they are expected to repress their nature with glass, barriers, and laws, a small band of resistors experiences an inexplicable but altogether human deficit of reason. D-503 even falls in love, finding within himself a living, breathing, hormonally responsive individual who yearns for basic biological satisfaction.

Much as the One State may demand that people believe in the "great, divinely bounding wisdom" of barriers in general, and the Green Wall in particular, the reader is soon made aware of the horror that comes from abandoning natural, human tendencies. The "numbers" who inhabit the One State of *We* are "compelled" to be content, or at least, pleased with their "mathematically infallible happiness." As with *Brave New World* and *1984*, such happiness is supposed to come in large part from rational, logical, precise state control over sex and reproduction. The descent from *We* to *Brave New World* and *1984* is clear: Zamyatin described a system designed to regulate sex through "child-breeding," as akin to "poultry-breeding or fish-breeding," all in an attempt to keep reproduction from occurring "as often and as much as anyone might wish [...] like animals." Of course, the One State ignores a fundamental flaw in its glorious, überscientific plan: the "numbers" are, in fact, human beings. And also animals.

Just as people in normal life often encounter various *memento mori* (reminders of one's eventual death), *mementi animalum* pop up unavoidably in the One State and in D-503's psyche: "even in our time," in which biology is supposedly so controlled as to have been overcome, "the wild, ape-like echo still occasionally rises from somewhere below, from some shaggy depth." It is that shaggy depth that especially interests us, even as it disconcerts D-503. Our hero ends up feeling—to his surprise, but not the reader's—lust, love, and even sexual jealousy, even though the One State, as in *Brave New World*, proclaims a "*Lex Sexualis*" in which "each number has a right to any other number, as to a sexual commodity." D-503's animal nature insists on being a sexy, selfish individual, not just a number in a vast, logically structured, marvelously efficient insect colony. Indeed, the insect parallel is quite explicit in *We*—just replace "six-wheeled" in the following description with "six-legged": "Every morning, with six-wheeled precision, at the same hour and the same moment, we—millions of us—get up as one. At the same hour, in million-headed unison, we start work; and in million-headed unison we end it. . . ."

Since people are mammals, not colonial insects, it is dystopian in the extreme to squeeze human beings into a beehive or an anthill. One way to deal with such deformation of human needs is to suffer: witness *We*. Another is to laugh. Which brings up *Antz*, an animated movie that begins with a hilarious scene in which Z, a troubled ant, is speaking (with the voice of Woody Allen) to a therapist about his feelings of "insignificance." The therapist approves enthusiastically: "Being an ant is being able to say, 'Hey—I'm meaningless, you're meaningless . . . let's be the best superorganism we can be!'" The reality is that the best superorganism a human being can be is a terrible superorganism indeed—or at least a terribly unhappy human being, one whose enforced "We" is unlikely ever to be reconciled with the biological "me."

Despite the inherently depressing plotlines of most dystopias, they remain persistently popular. *The Handmaid's Tale*, a modern feminist classic by Margaret Atwood, warns—like so many dystopias—of a future in which "love is not the point." And neither, of course, is motherhood or child-rearing. Ironically, this novel was actually intended as a criticism of evolutionary thinking, which Atwood interprets as oppressing women by enshrining reproduction as their sole biological and cultural "role." Notwithstanding her distrust of sociobiology, it is Atwood's paradoxically acute grasp of evolutionary realities—especially the centrality of reproduction—that makes *The Handmaid's Tale*, as well as her more recent work, *Oryx and Crake*, such a powerful dystopian story.

Part of human biology is, surprisingly for some, a yearning for culture. Although it might seem that biology and culture are antithetical, a capacity for culture is in fact one of humanity's most firmly established biological traits. It is thus notable that most literary dystopias include a suppression of the arts and humanities generally, and of literature in particular. Women are forbidden to read in *The Handmaid's Tale*; literature is disdained in *Brave New World*. Language is cynically perverted in *1984*; and the humanities don't even exist in the world of

We. In one of the best-known imagined dystopias, Ray Bradbury's *Fahrenheit 451*, the job of "firemen" is to *set* fires, not put them out—specifically, to burn books, and the cultural life they contain.

Just as *Fahrenheit 451* depicts a world in which cheap, artificial entertainment substitutes for the "real thing," the phenomenally popular movie *The Matrix* describes a vision that is even more nightmarish: a computer-generated cyberworld in which human beings, deceived as to their true situation, believe that they are living genuine lives. But they aren't. Most are victimized by a vast network of machines, their bodies preyed upon while their minds wander, misled, in a virtual "matrix" that is literally drained of its organicity. *The Matrix* is thus a prime example of a life-denying, biology-perverting dystopia.

Missing from *The Matrix* are any life-forms besides human beings, excepting a few birds and one cat that keeps reappearing, signifying a hacker intruding into that phony but satisfied cyberworld. But the cat that inhabits *The Matrix* isn't a purring, demanding, self-gratifying cat. Not a proud cat, or a Cat with a Secret Name. There *is* a cat, but it is a flat cat, not a fat cat; it lacks essence of Feline. To paraphrase T. S. Eliot, it does not have every cat's "ineffable effable/ Effanineffable/ deep and inscrutable singular Name."

Outside the movie studio, in the real and wondrous organic world of biology, there is indeed a cat. And a dog, and a hippopotamus, and a maple tree. And there is indeed a code, too, but it is written in base 4, not computer binary, and it consists of four letters: A, C, T, G.*

Look around you, decode the real world, and you may see the streams of oxygen and hydrogen, sulfur, and phosphorus. You may imagine uranium and plutonium, and all the busy, buzzing little atoms that burst from cyclotrons, born but to die in nanoseconds. It is a wonderful Zen exercise to look at a spoon and see iron, tin, copper, and even protons and electrons. However, the programs that will make you laugh and cry, love and grieve are written in purines and pyrim-

* I herein omit U, uracil, not because it is unimportant, but because the focus is on DNA, not RNA at this time. To be entirely fair, poor U should not be discounted, nor RNA, nor mitochondrial inheritance, etc., but for the sake of poetry I shall discuss only the Big Four.

idines, adenine, cytosine, thymine, and guanine. While *The Matrix* gives an illusion that protein gruel is really champagne and steak, ACTG gives us champagne and steak. The real thing.

And most people like champagne and steak, or champignons and escargot, because of their (our) shared cytoplasmic fondness for substrates of human metabolism.

An article in the *New England Journal of Medicine* noted that 60 to 80 percent of human disease-causing genes have parallels in the fruit fly. Slightly fewer are found in the worm *C. elegans*, while the zebra fish *Danio rerio* has genes that are counterparts to almost every disease-causing gene in H*omo sapiens*. Moreover, the newly emerging science of comparative genomics reveals that where there are shared disease-causing genes, there are also shared "normal" ones. William Blake was aware of the congruence between flies and humans, even in the 18th century:

> Am not I
> A fly like thee?
> Or art not thou
> A man like me?

Spend a minute gazing out your window. If you live near Times Square (or in much of Chicago, Los Angeles, Philadelphia, Boston etc.), and you will likely see a universe rather like *The Matrix*, a city crowded with inanimate things and very busy people, traveling hither and thither, eating or wooing or begging or selling things, along with an occasional pet dog or stealthy pigeon. It appears, from Times Square, that reality is composed of *The New York Times*, neon, cement, and a hive of people engaged in constant sociality. At the same time, "The ordinary city-dweller," wrote philosopher Susanne Langer, in 1951

> knows nothing of the earth's productivity: he does not know the sunrise and rarely notices when the sun sets; ask him what phase the moon is in, or when the tide in the harbor is high, or even how high the average tide runs, and likely as not he cannot answer you. Seed time and harvest are nothing to him . . . he probably does not feel the

power of nature as a reality surrounding his life at all. His realities are the motors that run elevators, subway trains, and cars, the steady feed of water and gas through the mains and of electricity over the wires, the crates of food-stuff that arrive by night and are spread for his inspection before his day begins, the concrete and brick, bright steel and dingy woodwork that take the place of earth and waterside and sheltering roof for him. . . . Nature, as man has always known it, he knows no more.

The Matrix—movie and pop phenomenon—is long gone, but "the matrix"—dystopic substitution/replacement of the unreal for the real—is altogether current, in what increasingly passes for reality as well as in stories.

Plato's "allegory of the cave" famously suggested that people perceive merely a simulacrum of reality, shadows thrown upon a cave's wall, artifice and deception instead of reality. The image still resonates, notably in the work of Portuguese Nobel laureate Jose Saramago, whose novel *The Cave* creates yet another dark, haunting, dystopian vision of the growing artificiality and sterility of modern life. In it, a small family is forced to migrate into a vast, arid, life-denying complex, called "the Center"—a "matrix" of sorts. At the end, the reader encounters the cave of the book's title and Plato's allegory, complete with mummies, chains, a wall, and evidence of fire. Saramago's dystopic message? It comes directly from Plato's *Republic*, also used as an epigraph by the author: "What a strange scene you describe and what strange prisoners. They are just like us."

Humanity, according to T. S. Eliot, cannot stand too much reality. In fact, the opposite seems more likely: humanity cannot stand too much *un*reality. And yet, unreality is precisely what people have been getting, in increasing amounts. More and more, people "experience" their lives vicariously via movies and spectator sports, do business and even "communicate" digitally, leading lives of quiet desperation that are increasingly removed from the organic reality, the green, oozy, smelly, breathing, slurping, biological matrix that was humanity's evolutionary cradle and for which and from which its DNA code has emerged.

It should occasion no surprise that people—increasingly deprived of their genuine matrix—resonated with the theme of the movie *Matrix*, that something is desperately wrong. Maybe we aren't really living what passes for our lives at all, but have somehow been inserted into a gigantic cyberworld in which the widespread, creeping sense of unreality is matched by an even creepier possible reality: that we aren't experiencing reality at all!

By one definition, a matrix is "that which gives origin or form to a thing," or "the basic substance in which particular items are embedded." The word originates from the Latin *mater*, or mother, and in a genuine sense, the organic soup (later, the complex organic world) in which human beings evolved is the mother of us all. It is both ancient and current, and its nourishing qualities are reaffirmed by modern human beings whenever they walk barefoot on the grass, or affiliate with an animal.

There are no pets in *The Matrix*'s fictional "Zion," which represents the last outpost of humanity struggling to be "genuine." Yet there are lots of pets kept by human beings in the real world. Why do people keep pets? After all, from a strictly Darwinian standpoint there is no payoff to pet keeping. Human beings are more related to one another than to their pets, yet they lavish time, resources, and love on these more distant genetic relatives, not a clever Darwinian move. Pets may be useful for soliciting mates (nothing like a Golden Retriever to make a person seem loved, loving, and thus, lovable), or genuine work (hunting drugs, guarding sheep), but the utility function of a pet in this regard seems a lot less than the cost of vet bills and cleaning expenses. An evolutionary calculus would insist that pets give back to their owners something that is worth their upkeep, or else there is inadequate payoff for the altruist, the owner. What is the adaptive significance of the Shih Tzu? Miniature horses? Persian cats? Why on earth would anyone shell out the money needed to maintain a Percheron, a Leonberger, or a Sulfur-crested Cockatoo?

Perhaps animals represent, once again, the last outpost of humanity struggling to be genuine. How else explain the fervent insistence by so many people to keep animals, even in the most difficult circumstances? And not only that, to touch them, care for them, be touched by them, slurped by dogs, kneaded by cats, parasitized by horses, serenaded by birds, entertained by fish and turtles, and, as though that is not enough, to surround their children with teddy bears, Mickey mice, Big Birds, and friendly—if obnoxious—purple dinosaurs? E. O. Wilson calls it "biophilia," a deep-seated human instinct to connect with nature. I suspect it represents, as well, a profoundly necessary assertion, by human beings, of their human kindred with all other life: ACTG.

It is also, increasingly, a kind of rebellion.

There is nothing new in perceiving *Homo sapiens* as endangered by its own creations. Semiscientific visionaries, like H. G. Wells, foresaw a world in which the values of the machine depersonalized our species and ultimately destroyed it. *The Time Machine* painted a world in which romantic and humanistic values were embodied in the child-like and helpless Eloi—late-19th-century flower children—who were preyed upon by the cruel, rapacious Morlocks, machine-oriented troglodytes who ate human flesh; in Wells's fantasy, the Morlocks no less than the Eloi were the victims of technology, condemned to a brutal and cheerless underground existence as slaves to their own machines.

All this is contrary to Aristotle's cheerful prediction: "If every instrument, at command, or from foreknowledge of it's master's will, could accomplish its special work . . . if the shuttle thus would weave and the lyre play of itself, then neither would the chief workman want assistants nor the master slaves." Although machines have in fact conferred many advantages, they have not rendered slavery obsolete; in fact, the cotton gin actually increased the demand for slaves in the American South, and cybernetics guru Norbert Wiener asked "May not man himself become a sort of parasite upon the machine? An affectionate machine-tickling aphid?" Or maybe not so affectionate.

In *We*, Zamyatin depicted a human anthill, a sterile, artificial, technology-driven world in which messy biology has been banished in favor of efficient artifice, and nature exists only on the far side of a man-made wall. The reader learns that "man ceased to be a savage only when we had built the Green Wall, when we had isolated our perfect mechanical world from the irrational, hideous world of trees, birds, animals." But even here, the thrumming of organicity can still be heard, and D-503 eventually sees, through the barrier separating his own sanitized, inhuman society from the messy world of organic nature, "the blunt snout of some beast star[ing] dully, mistily" at him, whereupon he is left with a not-so-surprising insight about the Green Wall: outside is more real, more whole, more natural, than in. Contemplating the beast on the other side, he asks himself, and the reader, whether "this yellow-eyed creature, in his disorderly, filthy mound of leaves, in his uncomputed life, is happier than we are?" and the answer is clear: Of course.

Nor is there much happiness in *The Matrix*, precisely because it takes place on the wrong side of a computer-generated Green Wall, imagined by the Warshowski brothers and embraced—with a knowing shiver—by millions of moviegoers.

"Only connect," wrote E. M. Forster. But despite the fears of the most technophobic, and despite the often unspoken alienation that made *The Matrix* such a mega-hit, connected we already are. Just take a good close look at a dog, or yourself, at a horse, or a worm, at a fly, or a cat gctgactgcatcatcggttc . . .

It should occasion no surprise that 21st-century audiences—increasingly deprived of their genuine matrix—have resonated with the warnings of dystopian storytelling. In one of *Brave New World*'s more memorable lines, the Controller asks John (and presumably also the reader): "So you don't much like civilization, Mr. Savage?" Once again, the answer to this simple question is clear: The stubborn savage living within each of us feels desperately out of place when we become, as A. E. Housman put it, "a stranger and afraid, in a world I never made."

"Something there is," wrote Robert Frost, "that doesn't love a wall." Something there also is, within each of us, that hates any hint of a wall between our innermost biological selves and the lives we may be forced to lead. When people are expected, or even imagined, to be thus separated from their biology—from themselves, in the deepest sense—dystopia follows.

21

Evolution's Odd Couple

IT HAD BEEN THE WORLD'S FIRST MURDER. THE APE-MAN EXULTANTLY threw his club (actually, the leg bone of a zebra) into the air, and as it spun, it morphed into an orbiting space station. In this stunning image from Stanley Kubrick's *2001: A Space Odyssey*, millions of moviegoers saw the human dilemma in microcosm. We are unmistakably animals, yet we also behave in ways that transcend the merely organic. Ape-men all, we are the products of biological evolution—a Darwinian process that is both slow and altogether organic—yet at the same time we are enmeshed in our own cultural evolution, which, by contrast, is blindingly fast and which proceeds under its own rules.

We are on Mr. Toad's wild ride, yet at the same time, we are all toads, perfectly good biological critters who aren't prepared even to drive "motor-cars," not to mention nuclear weapons. Therein lies the rub: Much of the human dilemma derives from our peculiar existence, simultaneously, in two worlds, the often inconsistent realms of biology and of culture. If dystopias derive from a fundamental disconnect between our biological, human nature and the lives people make for themselves (or, in the case of dystopian literature, what they imagine might be made) to some extent human beings all occupy an ongoing dystopia, thanks to the disparity between biological and cultural evolution.

While the cinematic ape-man's club traveled through air and, ultimately, into outer space, director Stanley Kubrick collapsed millions of years of biological and cultural evolution into five seconds. My point, however, is that this isn't simply a cinematic trick. We are all time-travelers, with one foot thrust into the cultural present and the other stuck in the biological past. And although it seems arrogant to propose anything as THE root of our human difficulty, I am about to do just that.

"It is dangerous," wrote Pascal,

> to show man too clearly how much he resembles the beast without at the same time showing him his greatness. It is also dangerous to allow him too clear a vision of his greatness without his baseness. It is even more dangerous to leave him in ignorance of both. But it is very profitable to show him both.

As purely biological creatures, we are neither more nor less "great" than our fellow organic beings. As cultural creatures, however, we are extraordinary indeed. We compose symphonies, travel to the moon, and explore the world of subatomic particles. But at the same time we are unique among living things in being genuinely uncomfortable in our situation. This should not be surprising, because even though our cultural greatness must have somehow derived from our organic beastliness, the two processes (organic and cultural) have become largely disconnected; and as a result, so have we: from ourselves, each other, our environment.

The little hyphen in ape-man is the longest line imaginable, connecting two radically different worlds, one biological and one cultural. Imagine two people chained together; one a world-class sprinter, the other barely able to hobble. Now, imagine that they are both expected to run as quickly as possible: The likely outcome includes a bit of tension all around.

To understand why biological and cultural evolution can experience such conflict (despite the fact that they both emanate from the same creature), consider the extraordinarily different rates at which they proceed. Biological evolution is unavoidably slow. Individuals, after all, cannot evolve. Only populations or lineages do so. And they are shackled to the realities of genetics and reproduction, since organic evolution is neither more nor less than a process whereby gene frequencies change over time. It is a Darwinian event in which new genes and gene combinations are evaluated against existing options, with the most favorable making a statistically greater contribution in succeeding generations. Accordingly, many generations are required for even the smallest evolutionary step.

By contrast, cultural evolution is essentially Lamarckian, and astoundingly rapid. Acquired characteristics can be "inherited" in hours or days, before being passed along to other individuals, then modified yet again and passed along yet more—or dropped altogether—everything proceeding in much less time than a single generation. Take the computer revolution. In just a decade or so (less than an instant in biological evolutionary time), personal computers were developed and proliferated (also modified, many times over), such that they are now part of the repertoire of most technologically literate people. If, instead, computers had "evolved" by biological means, as a favorable mutation to be possibly selected in one or even a handful of individuals, there would currently be only a dozen or so computer users instead of a billion.

Just a superficial glance at human history shows that the pace of cultural change has been not only rapid—compared with the rate of biological change—but if anything the rate of increase in that change seems itself to be increasing, generating a kind of logarithmic curve. Today's world is vastly different from that of a century ago, which is almost unimaginably different from 50,000 years ago . . . not because the world itself has changed, or the biological nature of human beings, but because such cultural creations as fire, the wheel, metals, writing, printing, electricity, internal combustion engines, television, and nuclear energy have been generated at blinding speed.

Try the following Gedanken experiment. Imagine that you could exchange a newborn baby from the mid-Pleistocene—say, 50,000 years ago—with a 21st-century newborn. Both children—the one fast-forwarded no less than the other brought back in time—would doubtless grow up to be normal members of their society, essentially indistinguishable from their colleagues who had been naturally born into it. A Cro-Magnon infant, having grown up in 21st-century America, could very well end up subscribing to *The Chronicle of Higher Education*, while the offspring of today's professoriate would fit perfectly into a world of mastodon skins and chipped stone axes. But switch a modern human adult and an adult from the late Ice Age, and there would be Big Trouble, either way. Human biology has scarcely changed in tens of thousands of years, whereas our culture has changed radically.

Admittedly, our capacity for culture is itself a product of our biological evolution, and yet, this is no guarantee that the two must proceed in synchrony. If anything, the opposite is more likely, since culture, like an errant and headstrong child—or Frankenstein's monster—has become disconnected from its biological moorings and has pretty much developed a momentum of its own, proceeding nearly independently of the biological process that originally spawned it. This is because cultural evolution has the capacity to take off on its own, to mutate, reproduce, and spread with such speed as to leave its biological parent far behind in the dust. In theory, the two might still be pointed toward the same ends, but biological evolution remains shackled by genetics—and thus, it lumbers along at the pace of a tortoise, never faster than one generation at a time, and nearly always much slower than that—while cultural evolution plays by its own rules, which often means a mad dash, like a hare. There isn't even much reason to expect the two to be headed in the same direction.

In Aesop's fable, the tortoise eventually wins, because the hare is foolish, overconfident, and easily distracted, whereas the tortoise (although slow) is persistent. In the real world, culture and biology differ in speed but they are equally foolish and equally stubborn. Most important, they will both cross the finish line together because despite their differences they are inextricably tied to each other. It is a strange spectacle, a cosmic sack race featuring two mismatched Siamese twins. Except that we are part of the show.

Don't misunderstand: Cultural evolution isn't altogether foreign to certain animal species (doubters might want to consult Frans de Waal's book *The Ape and the Sushi Master*), and biological evolution leaves a definite imprint on human behavior (much of my own writing and research, such as *Revolutionary Biology* and *The Myth of Monogamy*, seeks to make precisely this point). But it remains true that in the case of human beings, culture provides the dominant context while biology lurks in the background, with the two intersecting and interacting in ways that suggest few generalizations except this: The outcome is likely to be troublesome.

Fortunately, there can also be considerable harmony between our culture and our biology, in part because our biology is flexible, like a "one-size-fits-all" garment, able to conform to many different shapes, and in part because culture isn't so stupid as to attempt to force our biology to conform to patterns that are "inhuman," such as a society in which people are expected to sleep 23 hours a day, or not at all. Whereas all human behavior derives from both biology and culture, both nature and nurture, it does not necessarily follow that biology and culture are always comfortably adjusted to each other. On balance, just as we must look to the interaction between nature and nurture for the sources of our behavior, we can look to the conflict between nature and nurture for the sources of most of our difficulties. A useful rule in murder mysteries has long been "cherchez la femme"; when *Homo sapiens* is having trouble, a useful rule—although not yet a cliché—would be to look for possible conflict between the hare and the tortoise.

To change the metaphor: Two huge continents have drifted apart and now these great tectonic plates, culture and biology, grind together. The results, as we shall see, range from nearly trivial squeaks and wriggles, such as our troublesome sweet tooth or some of our sexual peccadilloes, to the most portentous quakes, including nuclear war, environmental abuse, and overpopulation, while in between lie a host of middle-sized tremors such as personal alienation and family dysfunction. The conflict between culture and biology, the Siamese sack race between hare and tortoise, is an event of paradoxical proportions, ranging from the seismic to the microscopic, from whole societies (indeed, the whole planet and its past, present, and future) to individual people and their likes and dislikes.

It has been said that when your only tool is a hammer, everything looks like a nail. Anyone trying to decipher the origins of human distress should be equipped with many tools, of which an appreciation of the conflict between biological and cultural evolution is but one. As hammers go, however, it seems especially useful, and looking around at the world, it is hard not to see a great many nails.

Here are a few examples. Let's start with violence and aggression, since this, after all, was what our cinematic ape-man was doing when so adroitly captured on film. The history of "civilization" is, in large part, one of ever-greater efficiency in killing: with increasing ease, at longer distances, and in larger numbers. Just consider the "progression" from club, knife, and spear to bow and arrow, musket, rifle, cannon, machine gun, battleship, bomber, and nuclear-tipped ICBM. At the same time, the human being who creates and manipulates these marvelous devices has not changed at all. Considered as a biological creature, in fact, *Homo sapiens* is poorly adapted to kill: the reality is that with our puny nails, nonprognathic jaws, and laughably tiny teeth, a human being armed only with his biology is hard-pressed to kill just one fellow human, not to mention hundreds or millions. But culture has made this not only possible but easy.

Animals whose biological equipment makes them capable of killing each other are generally disinclined to do so. Eagles, wolves, lions, and crocodiles have been outfitted by organic evolution with lethal weaponry and, not coincidentally, they have also been provided with inhibitions to their use against fellow species members. (This generalization was exaggerated in the past. Today, we know that lethal disputes, infanticide, and so forth do occur, but the basic pattern still holds: Rattlesnakes, for example, are not immune to each other's venom, yet when they fight, they strive to push each other over backward, not to kill.) Since we were not equipped, by biological evolution, with lethal weaponry, there was little pressure to balance our nonexistent organic armamentarium with behavioral inhibitions concerning its use. One reason why guns are so dangerous is that the lethal consequence of a very small movement—curling a finger around a trigger, with barely a few ounces of pressure—are magnified, by superb technology, into violent acts of dreadful consequence. If, by contrast, we had to live—and die—by the application of direct, biological force alone, there would be far more living and less dying.

The dreadful history of human-human slaughter also leaves little doubt that our species is not automatically prone to respect the subordination gestures of potential victims. Moreover, even if he possessed such instinctive inhibitions on his own lethal violence (and as we have seen, it usually is *his* violence), the bombardier, flying 20,000 feet above his victim—or the military leader on a distant continent with his finger on The Button—couldn't even perceive such gestures; assuming, of course, that he were predisposed to respond to them.

The disconnect between culture and biology is especially acute in the realm of nuclear weapons. At the one-year anniversary of the bombing of Hiroshima, Albert Einstein famously noted that "the splitting of the atom has changed everything but our way of thinking; hence we drift toward unparalleled catastrophe."

He might have been talking about musk oxen. These great beasts, like shaggy bison occupying the Arctic tundra, have long employed a very effective strategy when confronted by their enemies: wolves. Musk oxen respond to a wolf attack by herding the juveniles into the center, while the adults face outward, arrayed like the spokes of a wheel. Even the hungriest wolf finds it intimidating to confront a wall of sharp horns and bony foreheads, backed by a thousand pounds of angry pot roast. For countless generations, their antipredator response served the musk ox well. But now, the primary danger to musk oxen is not wolves, but human hunters, riding snowmobiles and carrying high-powered hunting rifles. Under this circumstance, musk oxen would be best served if they spread out and hightailed it toward the horizon, but instead they respond as previous generations have always done: They form their trusted defensive circle, and are easily slaughtered.

The inventions of the snowmobile and the rifle have changed everything but the musk ox way of thinking; hence they drift toward unparalleled catastrophe. (Musk oxen are currently a threatened species.) They cling to their biology, even though culture—our culture—has changed the payoffs. Human beings also cling to (or remain unconsciously influenced by) their biology, even as our own culture has dramatically revised the payoffs for ourselves as well. That musk ox–like stubbornness is especially evident when it comes to thinking—or not thinking—about nuclear weapons.

Take, for example, the widespread difficulty so many people have when it comes to conceiving nuclear effects. When told something is "hot," human beings readily think in terms of boiling water, burning wood, or perhaps molten lava. But the biological creature that is *Homo sapiens* literally cannot conceive of temperatures in the millions of degrees. Before the artificial splitting of uranium and plutonium atoms (a cultural/technological innovation if ever there was one), nuclear energy had never been released on earth. No wonder we are unprepared to "wrap our minds" around it. Similarly with the vast scale of nuclear destruction: We can imagine a small number of deaths—so long as none of them include our own!—but are literally unable to grasp the meaning of deaths in the millions, all potentially occurring within minutes. And so, ironically, the conflict between our biological natures and our cultural products has in itself cloaked nuclear weapons in a kind of psychological untouchability.

By the same token, the "caveman" within us has long prospered by paying attention to threats that are discernible—a stampeding mastodon, another Neanderthal with an upraised club, a nearby volcano—while remaining at the same time less concerned about what cannot be readily perceived. Since nuclear weapons generally cannot be seen, touched, heard, or smelled, they tend to evade our radar, allowing the nuclear Neanderthal to function as though these threats to his and her existence don't exist at all. (Not surprisingly, national policies that "refuse to confirm or deny" the presence of such weapons add further yet to their aura of invisibility, and hence, nonexistence.) If a homicidal lunatic were to stalk your workplace, or if a fire suddenly broke out, you would doubtless respond, and quickly. But although we are all stalked by a far more dangerous nuclear menace, the Neanderthal within us remains complacent.

According to Greek mythology, the gods punished Prometheus—who had impudently given fire to human beings—by chaining him to a great mountain, whereupon he was visited daily by a vulture, who chewed on his liver. Modern human beings, biological creatures acting not in deliberate evolutionary time but in a cultural frenzy, have unleashed a much more dangerous fire than Prometheus could ever have imagined, a fire made all the more lethal by the fact that, deep inside, we really aren't very modern ourselves. In *Prometheus Bound*, Aeschylus asks:

> Prometheus, Prometheus, hanging upon Caucasus,
> Look upon the visage of yonder vulture:
> Is it not thy face, Prometheus?

Two thousand years later, Pogo said it more simply: "We have met the enemy and he is us."

"From now on it will no longer be enough to ask if man can do something," wrote David Brower. "We must also ask whether he ought to." And so we come to our environmental crisis.

The reality is that living things do sometimes destroy their own environments. Elephants overgraze several national parks in Africa, foxes have been known to kill hundreds of gulls in a single night, and algal blooms—themselves "natural" phenomena—can make water toxic. But nearly always, the damage is limited in geographic extent and also in the number of species and of individuals involved. (Moreover, at least in recent years, the hand of *Homo sapiens* can often be discerned whenever large-scale bio-damage threatens.) As a general rule, living things, left to their own biological equipment, are simply unable to do very much damage. Human cultural evolution, on the other hand, has changed all that; much like the trigger of a gun, our cultural advances have served as an immense multiplier. As a result, we "can do" things of incredible destructiveness, including—but not limited to—the extinction of whole species, the pollution of continents, the damming (or as John Muir called it, the "damning") of rivers, the draining of aquifers, depletion of nonrenewable resources, and the lethal reconfiguring of those basic geo-thermo-chemical cycles on which planetary life depends. Other living things, lacking such capacities, have no reason to debate whether they ought to act upon them. Were it not for our cultural evolution, we, too, would be limited by what we are biologically capable of doing (which is to say, not much), rather than what we *should* do, given our unique capability of wreaking havoc.

Then there is the matter of population. Living things have been selected to reproduce at what is essentially their maximum rate, or rather, at a pace that results in projecting the maximum number of their genes into the future. Human beings are no different. Under "precultural" conditions, a high birth rate was offset by a high death rate, but one of humanity's proudest (cultural) achievements has been death control, mediated especially by public health measures, notably vaccinations, antibiotics, and advances in infant nutrition. Comparable culturally mediated mechanisms of birth control—contraceptive pills, condoms, IUDs—although available, have frequently been stymied by countervailing conservative religious ideology ("be fruitful and multiply," God is reported to have enjoined Adam). At the same time, our biologically given tendencies for full-throttle reproduction have not been significantly altered. The result? A rising level, just as when a sink's drain is plugged, but the faucets remain open.

Under strictly biological conditions, different species can be imagined pushing as hard as they can against each other, and similarly for individuals within each species. With everyone pushing hard, anyone who lets up loses out. But thanks to cultural evolution, we have eliminated much of the natural resistance that our own potentially expanding numbers would otherwise have encountered. As a result, there is comparatively little at the moment for us to push against, so that if we don't ease up—which is to say, wise up—we must eventually fall on our faces.

The troublesome combination of biology and culture reveals itself in many dimensions, the personal no less than the planetary. For a final interconnected example, consider obesity, heart disease, and our species-wide sweet tooth. Most people like sugar. Why? Almost certainly because our primate ancestors were fruit-eaters, and ripe fruit has lots of sugar. That's why sugar is sweet; if we were ant-eaters, we'd doubtless exclaim over the sweetness of ants, and perhaps note in passing that ripe peaches are unpleasantly bitter. In any event, our primate sweet tooth served us well in a world of strictly biological evolution, where sugars were present in large quantities only in the intimate company of healthy, fruity nutrition. But clever cultural creatures that we are, human beings have developed the confectionery industry and the ability to produce candy, chocolates, and soft drinks galore, laden with sugars and little else.

Another dietary consequence: Our Pleistocene ancestors were also occasional carnivores who almost certainly treasured the opportunity to eat meat when available, which wasn't often. Moreover, given that wild game is usually quite lean, the likelihood is that eating fat was a special treat, high in caloric value and encountered only on extraordinary occasions. It would be surprising, therefore, if our species didn't evolve with a particular fondness for eating meat, and particularly fatty meat, when possible. Today, it is quite possible. Even moderately well-to-do consumers can avail themselves of "well-marbled" steaks and greasy fast foods, comestibles whose appeal is almost certainly due to a biologically generated fondness for something that throughout most of our evolution wasn't available and when it was, beckoned as a rare and highly desirable opportunity.

Add to this another likely consequence of the biology/culture disconnect, and our difficulties are exaggerated. Thus, our ancient ancestors weren't couch potatoes. They had to walk, run, and otherwise exert themselves. By contrast, much of cultural evolution has involved "labor-saving devices" such as automobiles, elevators, telephones, and television monitors, enabling us to avoid the expenditure of calories. It makes sense that our Pleistocene forebears would jump at the chance to be as inert as possible; that is, to avoid jumping (or running, walking, and so forth) whenever they could get away with it. Our body-weight regulating system, not to mention our cardiovascular system, would therefore have evolved in a context of unavoidable physical activity combined with a low-fat, low-sugar diet. Thanks to cultural evolution, modern human beings, by contrast, are able to indulge a biological fondness for high-fat, high-sugar diets, and at the same time, they must remind themselves to go out of their way to obtain exercise.

If you are "with me" at this point, you might be eager for some proposed solutions, but here, alas, the biology/culture interface yields few reliable insights. One thing, however, seems clear: Biology will not ride to our rescue. Evolution by natural selection is simply too slow, too shortsighted and stupid, too unresponsive to the challenges generated by our concurrent cultural evolution. Insofar as our difficulties derive from the disconnect between human biology and culture, we must look, therefore, to the latter, and to our large and multicompetent brains, for ways out of our current mess. After all, the remarkable flexibility of our cognitive capacities is itself a biological trait, by which evolution has endowed us, paradoxically, with the ability to transcend our own proclivities. We, alone among all living things, can say "No" to many of our inclinations, shed light on our blind spots, and maybe even help reconcile our ancient biology with our modern culture, to the ultimate benefit of ourselves, our fellow creatures, our shared planet.

In his *Second Discourse*, Jean-Jacques Rousseau argued that human beings were ill-served by society and civilization, that people had been nobler, purer, and altogether superior creatures when in a "state of nature." He sent a copy of this essay to Voltaire, who wrote back:

> I have received, Monsieur, your new book against the human race. . . . You paint in very true colors the horrors of human society . . . no one has ever employed so much intellect to persuade men to be beasts. In reading your work one is seized with a desire to walk on four paws. However, as it is more than sixty years since I lost that habit, I feel, unfortunately, that it is impossible for me to resume it."

I am not about to cross swords with Voltaire: we cannot simply toss out our cultural advances. Despite the difficulty of being so dependent on culture, we have gone too far down this road to turn back.

> "Would you tell me, please, which way I ought to go from here?," Alice, lost in Wonderland, asked the Cheshire Cat.
>
> "That depends a good deal on where you want to get to," said the Cat.
>
> "I don't much care where—" said Alice.
>
> "Then it doesn't matter which way you go," said the Cat.
>
> "—so long as I get somewhere," Alice added as an explanation.
>
> "Oh you're sure to do that," said the Cat, "if you only walk long enough."

It is "natural" for a species with a well-developed culture to tamper with its biology, and unnatural to refrain, just as it is "natural" for a species with our particular biology to develop the kinds of cultural products that so delight and bedevil us today. British scientist Dennis Gabor once suggested that our job was to invent the future. One way or the other, as the Cheshire Cat says, we are sure to do that. We may not know who is in the saddle, who is riding whom, or exactly where we are going, but we're certainly on our way.

Index

abortion, 32–35
ACTG, 167–68, 171
Adams, Douglas, 13, 15
adultery, avian, 159
Aeschylus, 180
Aesop, 176
African Queen, The (Forester), 101
African Queen, The (movie), 101
aggression, 114, 142, 147, 148, 154, 177–79. *See also* violence
AIDS (Acquired Immune Deficiency Syndrome), 90
Aldrich, Henry, 72
Alex (parrot), 68
Alex Papers, The (Pepperberg), 68
Alice in Wonderland (Carroll), 184
Allen, Woody, 166
altruism, 47, 81–83, 84, 92, 102, 105, 106–35
amity, in-group, 132
animal rights, 35–36
anthropomorphism, 69
ants, 43, 82, 152–53
Antz (movie), 166
Ape and the Sushi Master, The (de Waal), 176
apes, 53–54, 89, 97, 137. *See also* primates; *specific apes*
Aristotle, 13, 71, 171
Asperger's Syndrome, 83–84
Astonishing Hypothesis, The (Crick), 49
Atwood, Margaret, 166
autism, 83–84
babies, 45–46
Bacon, Francis, 11
bacteria
  cholera and, 45
  drug resistance in, 29, 36, 38
  plague and, 43–44
bad faith, 92
barn swallows, 159
Bateson, William, 39
Becker, Ernest, 59
bees, 82, 156
beetles, longicorn Prionid, 39
behaviorism, 49–56
Behavior of Organisms, The (Skinner), 50
belief, 80–85
Bell Curve, The (Murray and Herrnstein), 64
Benjamin, Walter, 30, 38
Beyond Free Will and Dignity (Skinner), 50
Bible, 14. *See also* Daniel, book of; Genesis, book of; Old Testament
Bierce, Ambrose, 57
biophilia, 171
birds. *See also specific birds*
  adultery by, 159
  reproduction of, 138
birth control, 182
Bisson, Terry, 55–56
Black Death, 43–44
Blake, William, 168
Blind Watchmaker, The (Dawkins), 23
Bloom, Paul, 67

body, bodies
altruism and, 109, 127
genes and, 47, 128–29
Haig on, 110
mind and, 55
reproduction and, 45–46, 60–61, 105, 128
Bogart, Humphrey, 101
Bradbury, Ray, 167
Brahe, Tycho, 10–11, 12, 14
brain. *See also* mind
animal, 70
of an octupus, 111
Bisson on, 56
genesis of, 34–35
hypertrophy of, 111–12
intelligence and, 64
reproduction and, 60–61, 77
Brave New World (Huxley), 162–63, 164, 165, 166, 172
Brecht, Bertolt, 106, 108
breeding. *See* reproduction
Bridgewater Treatises, 16
Brower, David, 181
Bruyère, Jean de la, 108
Buchsbaum, Ralph, 42
Bundy, Ted, 137
Burns, Robert, 61
Bush, George W., 21, 64, 119

Cain, 150
Calvin, John, 149
camouflage, 157
Campbell, Anne, 147
Camus, Albert, 92, 94, 95
capitalism, 124
Carroll, Lewis, 49
caterpillars, 130
causation, 51–53, 55, 58
Cave, The (Saramago), 169
centrality, human, 9–12, 13–15, 109–10
change, 175–76
characteristics, inheritance of acquired, 39
cheating, 110, 115, 120, 123, 159–60
cheetahs, 159–60
chimpanzees, 35–36, 37–38, 66–67, 100, 137, 151–52
choice, 87, 94
cholera, 44–45
Churchill, Winston, 21
Civilization and Its Discontents (Freud), 30, 106, 107
Cleopatra, 74
Clever Hans (horse), 69
clones, 36
Columbine High School, shootings at, 53–54
communication, 155–61
communism, 121
computers, personal, 175
conception, 32–35
Conrad, Joseph, 148
consciousness, 49, 54, 55, 56, 57–63
constructionism, 85
Copernicus, Nicolaus, 9, 10, 14
Cosmides, Leda, 78–79
coughing, 44
Crane, Stephen, 71, 74–75
creation, special, 16
Crick, Francis, 49, 55, 57

D-503 (fictional character), 165–66, 171–72
Dahmer, Jeffrey, 137
Daly, Martin, 144, 145
Daniel, book of, 73
Dart, Raymond, 148
Darwin, Charles, 14
on altruism, 81, 82
Descent of Man by, 63
Freud on, 9
on historical continuity, 31
Huxley and, 101
Lyell's influence on, 38, 70
on natural selection, 38–39, 40
Nietzsche on, 96
Origin of Species by, 21, 54, 95
Dawkins, Richard, 23, 87, 105, 126, 158
de Beauvoir, Simone, 86
deconstructionism, 84
de la Mettrie, Julien Offray, 54–55

DeLay, Tom, 53–54
Dennett, Daniel, 21
Descartes, René, 55, 57, 66
Descent of Man, The (Darwin), 63
de Vries, Hugo, 39
de Waal, Frans, 153, 176
Diamond, Jared, 35–36
diarrhea, 44–45
differences, male-female, and violence, 136–47
dilemmas, social, 118–25
Dillard, Annie, 100
dinosaur, late-Cretaceous, 40
Discontinuity, Fallacy of, 32, 35–38
dishonesty, 155–61
disobedience, 93
disorders, behavioral, 143–44
Dostoyevsky, Fyodor, 59–60, 71, 73–74
dualism, 55
ducks, mallard, 122–23
dystopias, 162–72

Economy of Nature and the Evolution of Sex, The (Ghiselin), 131
Edelman, Gerald, 57
Egerton, Francis Henry, 16
Einstein, Albert, 57, 179
Eldridge, Niles, 40
elephants, 121–22, 139, 181
elephant seals, 139–40, 141, 145
Eliot, George, 102, 113
Eliot, T. S., 167, 169
embodiment, 55
emotions, in people and other life forms, 39
Emoto, Masaru, 58
Enduring Love (McEwan), 116
enmity, out-group, 132
Enquiry Concerning Human Understanding, An (Hume), 106
environmental crisis, 181
equilibrium, punctuated, 40
equipotentiality, 145
Essay on Man (Pope), 98
ethics, evolutionary, 90, 101–2, 104, 106, 112–14, 116, 134
Ethnographic Atlas (Murdock), 141
euthanasia, 35
evolution, cultural, 24–26, 173–84
exercise, 183
existentialism, evolutionary, 86–97
extinction, 40–41

Fahrenheit 451 (Bradbury), 167
faith, 75
fat, 182–83
Ferdinand the Bull (fictional character), 140
fertilization, 33–34
fetus, viability of, 32–35
Fisher, Ronald A., 39, 40, 130
fishes, 83
fleas, 43–44
Forester, C. S., 101
Forster, E. M., 172
foxes, 43, 181
free will, 50–53, 54, 87, 89, 94, 97, 110
Freud, Sigmund, 9, 30, 55, 93, 96, 106, 107, 108, 114
frogs, 83
Frost, Robert, 172
fruit fly, 168
Futuyma, Douglas, 13

Gabor, Dennis, 184
gain, personal, 118–25
Gama, King (fictional character), 105
game theory, 83, 125
Garber, Christian, 30–31
Gardner, Howard, 64
gazelles, 159–60
genes
    altruism and, 81–83, 132, 134
    behavior and, 53
    biological evolution of, 174–75, 176
    bodies and, 47, 128–29
    centrality and, 109–10
    consciousness and, 60–61
    Crick on, 49
    disobedience and, 93
    existentialism and, 87–88, 91, 92
    free will and, 47–48

Herrnstein on, 64
intelligence and, 64–66
love and, 52
murder and, 151
Murray on, 64
1984 and, 163–64
races and, 133
reproduction and, 22, 25, 45–46, 60–61, 99, 105, 121–22, 127
selfishness in, 62, 108–9, 110, 115, 117, 121–22, 126–31
self-perpetuation of, 100, 101
shared, 168
sociobiology and, 89, 95–97, 105
Genesis, book of, 37, 38
"Genetical Evolution of Social Behavior, The" (Hamilton), 127–28
genetics, coinage of the word, 39
genotype, and environment, 65
Geological Evidence of the Antiquity of Man, The (Lyell), 70
Ghiselin, Michael, 131
Ghost in the Machine, The (Koestler), 114
Gingerich, Owen, 81
Glendower, Owen, 157
global warming, 119
God
Calvin on, 149
consciousness from, 63
humans in image of, 96, 149
personal relationship with, 14–15
Golding, William, 148
Goldschmidt, Richard, 39, 40
Goodall, Jane, 38, 152
Gould, Stephen Jay, 40, 64
Goya, Francisco, 79
gradualism, naturalistic, 38–39, 40–41
Grahame, Kenneth, 65
Grant, Peter, 29–30
Grant, Rosemary, 29–30
Gravity's Rainbow (Pynchon), 111
Green, Ronald, 35
Gross, Paul R., 84
Gulliver (fictional character), 153–54
Gulliver's Travels (Swift), 72
Haig, David, 110
Haldane, J. B. S., 40, 130
Hamilton, William D., 81, 82, 105, 127–28, 129, 130, 131
Hamlet (fictional character), 23, 24, 71
handicap principle, 160
Handmaid's Tale, The (Atwood), 166
happiness, 171–72
Harcourt, Alexander, 153
Hardin, Garrett, 121, 124
Hardy, Thomas, 46
Harvey, William, 85
Heart of Darkness (Conrad), 148
Heidegger, Martin, 91
Henry IV (Shakespeare), 157
Hepburn, Katharine, 101
"Heredity" (Hardy), 46
Herrnstein, Richard, 64
Higher Superstition (Gross and Levitt), 84
High Tide in Tucson (Kingsolver), 135
Hitchhiker's Guide to the Galaxy (Adams), 13
HIV (Human Immunodeficiency Virus), 90
Hobbes, Thomas, 108, 123, 124, 125
Hölldobler, Bert, 152–53
Homicide (Daly and Wilson), 144
host manipulation, 43–48
Housman, A. E., 172
Hrdy, Sarah, 109, 150
humans, and other life forms, 35–38
Hume, David, 17, 53, 73, 90, 98, 106, 108
Hutton, James, 14
Huxley, Aldous, 149, 162–63
Huxley, Thomas, 21, 101, 102

inclusive fitness theory, 81–82, 93, 129, 130, 131, 183
inertia, 183
infanticide, 100, 150–51
influenza, 44
inheritance, Mendelian particulate, 40

intelligence, 64–70
intelligent design, 16–17, 21, 38

James, William, 149
Jaspers, Karl, 89
Jefferson, Thomas, 12
Johnson, Lyndon, 142
Kagan, Jerome, 102
Kahneman, Daniel, 76
Keats, John, 87
Kepler, Johannes, 10
Kermit the Frog (fictional character), 116
Khayyam, Omar, 71
Kierkegaard, Søren, 89, 91–92
killing. *See* murder
King, Stephen, 120
Kingsolver, Barbara, 135
kin selection, 47, 115, 117, 128–30, 132, 133–34
knowledge, forbidden, 104–10
Koestler, Arthur, 114
Krebs, John, 158
Krutch, Joseph Wood, 53
Kubrick, Stanley, 173
Kuhn, Thomas, 81
Kyoto Accords, 119

Lamarck, Jean-Baptiste, 39, 40
Langer, Susanne, 168
Leakey, Louis, 38
Lehrer, Tom, 156–57
Leibniz, Gottfried Wilhelm, 32
Lenin, Vladimir, 45
Leviathan (Hobbes), 123
Levitt, Norman, 84
liars, lying, 155–61
lice, 12
Life of Gladstone (Morley), 68
Lightman, Alan, 81
links, missing, 27–28
lizards, 159
Loeb, Jacques, 66
Lord of the Flies (Golding), 148
Lorenz, Konrad, 149–50, 151, 156
love, 52
Lucretius, 54
Lucy, 28
Lyell, Sir Charles, 38, 70

Machiavelli, Nicolò, 108
MacNeice, Louis, 23
Maggie's Dilemma, 113, 116
Malraux, André, 97
Man a Machine (de la Mettrie), 54–55
Man in the Modern Age (Kierkegaard), 89
manipulation, 157–58
Marriage of Figaro, The (Mozart), 103
Matrix, The (movie), 167, 168–69, 170, 172
Max Planck Institute for Evolutionary Anthropology, 66, 67
McEwan, Ian, 116
meat, red, 182–83
men, violent, 136–47
Men in Black (movie), 42
mental illness, 143
mice, 43
Middlemarch (Eliot), 102
Mill on the Floss, The (Eliot), 113
mind, 34–35, 54–55, 62, 70, 76–77. *See also* brain
Miranda (fictional character), 163
monkeys, 89, 122, 137, 142, 150–51, 158–59
monogamy, 140, 141, 142
monster, hopeful, 39
Montaigne, Michel de, 73
Moore, G. E., 90, 98
morality, 108–9, 116–25
Morgenthau, Hans, 142
Morley, John, 68
Morse, Samuel F. B., 126
Mozart, Amadeus, 103
Muir, John, 125, 181
murder, 131, 144–45, 146, 148, 149–50, 151–54, 177–79. *See also* infanticide
Murdock, George P., 141
Murray, Charles, 64
musk oxen, 179
mutations, mutation theory, 39–40

Myth of Monogamy, The (Barash), 176

Nash, Ogden, 98
natural, tyranny of the, 98–103
naturalistic fallacy, 90, 98, 110, 117
Natural Selection and Adaptation (Williams), 105
Natural Theology (Paley), 12
nature
  city dwellers and, 168–69
  and nurture, 177
  Rousseau on, 184
  Wilson on, 171
  Zamyatin on, 165–66, 171
nepotism, 47, 100, 115, 130, 131, 132
neurobiology, 58–59, 70
New England Journal of Medicine, 168
Newton, Sir Isaac, 32, 72–73, 75, 99, 127
New York Times, The, 120, 168
Nietzsche, Friedrich, 89, 92, 96, 107–8
1984 (Orwell), 162, 163–64, 165, 166–67
Notes from Underground (Dostoyevsky), 73
nuclear weapons, 179–80
nurture, nature and, 177

obedience, 93, 94
obligation, social, 112–17, 134
O'Brien (fictional character), 164
Old Testament, 14
Oliver (parrot), 69
Origin of Species, On the (Darwin), 21, 54, 95
Ortega y Gasset, José, 92
Orwell, George, 150, 162, 163, 164
Oryx and Crake (Atwood), 166

pain, 61
Paley, William, 12
parasites, 42–43, 65, 100, 160
Pascal, Blaise, 14, 72, 88, 89, 174
Pavlov, Ivan, 53
Pepperberg, Irene, 68
pets, 170–71
phenotype, 65
philosophy, existential, 86–87, 88, 89–91, 94. *See also* existentialism, evolutionary
Pilgrim at Tinker Creek (Dillard), 100
Pink Floyd, 104
plague, 43–44
Planck, Max, 99
Plant, The (King), 120
Plato, 71, 95, 96–97, 169
pleasure, 61
Poe, Edgar Allan, 80, 81
Pogo (fictional character), 181
polyandry, 141
polygyny, 140, 141–42
Pony Fish's Glow, The (Williams), 19
Pope, Alexander, 98–99, 109
postmodernism, 84–85
Post-Traumatic Stress Disorder (PTSD), 84
predation, predators, 100, 127, 130, 149, 151, 159. *See also* *specific predators*
primates, 97, 100, 137, 140, 153, 182. *See also* apes; *specific primates*
primroses, evening, 39
Princess Ida (Gilbert and Sullivan), 105
prions, 41
Prometheus Bound (Aeschylus), 180
propagation. *See* reproduction
Protagoras, 149
psychology, evolutionary, 104, 108–9. *See also* sociobiology
public good, 118–25
"Purloined Letter, The" (Poe), 80, 81
Pynchon, Thomas, 12, 111, 112, 113
Pythagoras, 72, 75

Quammen, David, 111

racism, 132–33, 134
rape, 100, 122–23, 137
rationality, 71–79, 97. *See also* reason
rats, 43–44
rattlesnakes, 178
reality, 169–70

reason, 71–79, 96–97, 109, 110
Reasons for Drinking (Aldrich), 72
reciprocity, 109, 115, 117, 122, 131
reinforcement, 51, 53
reproduction. *See also* conception
  altruism and, 82
  of birds, 138
  birth control and, 181–82
  bodies and, 45–46, 60–61, 105, 128
  brain and, 60–61
  in *Brave New World,* 163, 165
  differential, 25
  genes and, 22, 25, 45–46, 60–61, 99, 105, 121–22, 127
  in *The Handmaid's Tale,* 166
  in *Oryx and Crake,* 166
  selfishness and, 115, 121–22, 127
  violence and, 138–42
  in *We,* 164, 165
Republic, The (Plato), 169
retrorecognition, 81, 83
Revolutionary Biology (Barash), 176
Rico (dog), 67
Robertson, Pat, 90
Romantic movement, 73
Rousseau, Jean-Jacques, 73, 123–24, 184
Russell, Bertrand, 36, 108

Sandburg, Carl, 110
Saramago, Jose, 169
Sartre, Jean-Paul, 86, 87, 88, 92, 94
Science, 14, 67
science fiction, 162–72
Scott, Sir Walter, 153
Second Discourse (Rousseau), 184
selection, unit of, 126
selection differential, 29
Selfish Gene, The (Dawkins), 105, 126
selfishness, 112–35
  in genes, 62, 108–9, 110, 115, 117, 121–22, 126–31
  reproduction and, 115, 121–22, 127
  sociobiology and, 104, 108, 115
Sesame Street (TV program), 116
sex, 163, 164, 165, 166. *See also* reproduction
Shakespeare, William, 23, 24, 163
sheep, 43
Simon, Herbert, 76
Simpson, O. J., 137
Sisyphus, Myth of, 94, 95
Skinner, B. F., 49–56
Skutch, Alexander, 82
Smith, Adam, 124–25
Smith, John Maynard, 105
Smith, Susan, 137
Smith, Winston (fictional character), 163, 164
sneezing, 44
"Snow" (MacNeice), 23
Social Contract, The (Rousseau), 123–24
socialism, 121
sociality, elimination of, 164
social learning theory, 146–47
sociobiology
  Atwood's distrust of, 166
  creative arts and, 162–84
  ethics and, 116
  existentialism and, 86–97
  genes and, 89, 95–97, 105
  inclusive fitness theory and, 81–82
  selfishness and, 104, 108, 115
  teaching of, 108–9
Sociobiology (Wilson), 105
Socrates, 95–96, 104, 110
soul, the, 33, 34–35, 54, 63
South Pacific (show), 134
species
  benefit, 104–5
  endangered, 27
  evolution of, 21–31
  rejecting one's own, 150
  selection, 40
spite, 73–74
SpongeBob SquarePants (fictional character), 41
Stein, Gertrude, 66
stem cell research, 32–35
Stevenson, Robert Louis, 148
Stewart, Potter, 59, 155
Strange Case of Dr. Jekyll and Mr. Hyde, The (Stevenson), 148

strategies, evolutionarily stable, 83, 84
Strauss, Leo, 71
Structure of Scientific Revolutions, The (Kuhn), 81
sugar, 182
swallows, barn, 159
Swift, Jonathan, 44, 72, 153–54
tapeworm, 43
taxes, 120
technology, 154
Tempest, The (Shakespeare), 163
Tennyson, Lord Alfred, 154
terrorists, 146
testosterone, 143
thought, 96. *See also* consciousness
Threepenny Opera, The (Brecht), 106
Time Machine, The (Wells), 171
Tinbergen, Niko, 156
To a Louse (Burns), 61
tolerance, 134
"Tragedy of the Commons" (Hardin), 121, 124
Trivers, Robert L., 105, 109, 122
Tversky, Amos, 76
Twain, Mark, 27, 155
2001: A Space Odyssey (Kubrick), 59, 173

unconscious, Freud on, 9, 96
Underground Man (fictional character), 71, 73–74, 75, 79
uniformitarianism, 38
unreality, 169–70
urea, synthesis of, 41

variation, discontinuous, 39
violence, 114–15, 123, 131, 136–47, 177–79. *See also* aggression; *specific forms of violence*
viruses, 41, 44, 90
volleyball, 93–94, 95
Voltaire, 184
von Osten, William, 69

Wason Test, 77–78
water, shortage of, 119–20
Watson, John, 50, 54
Wealth of Nations, The (Smith), 124–25
weapons, of mass destruction, 154, 179–80
Wells, H. G., 171
West, Rebecca, 75–76
wetness, 60
We (Zamyatin), 164–66, 167, 171
White, Lynn, 14
Wiener, Norbert, 171
Williams, George C., 19, 90, 105
Wilson, Edward O., 105, 152–53, 171
Wilson, Margo, 144, 145
Wind in the Willows, The (Grahame), 65
Wolfgang, Marvin, 145, 146
wolves, 137, 149–50, 151, 152, 179
women, violence and, 137, 143–44
woodpeckers, 83
worms, 42–43
Wrangham, Richard, 152
Wright, Seawall, 40

Zahavi, Amotz, 160
Zamyatin, Yevgeny, 164–66, 171

DATE
DISCARDED BY
FREEPORT
MEMORIAL LIBRARY